Dairy Farming:
The Beautiful Way

Adam J. Klaus

Copyright © 2014 Adam J. Klaus

All rights reserved.

ISBN: 1502490773
ISBN-13: 978-1502490773

DEDICATION

To my three beautiful children, Liliana, Noah, and Marcel.
All born on Bella Farm, and all the inspiration and motivation I
needed to create this wonderful place in history.
You three are my guiding light, may you forever shine so bright.

And to Alison, mother of these three children, and tireless companion
creating Bella Farm. Nobody could have ever worked harder than you,
with more dedication, to make our farm dream a reality. May you always
have a gentle cow to milk on a peaceful summer day.

TABLE OF CONTENTS

1. THE BEAUTY OF DAIRY FARMING 1
2. REAL MILK 6
3. EQUIPMENT FOR THE PROFESSIONAL DAIRY 11
4. THE DETAILS OF MILKING 29
5. MILK HANDLING 47
6. ESTABLISHING THE DAIRY HERD 67
7. QUALITIES OF THE PERFECT COW 79
8. HOLISTIC HERD MANAGEMENT 91
9. PASTURE-BASED DAIRY NUTRITION 109
10. HAY, WINTER FEEDING, AND COW MINERALS 129
11. BREEDING AND CALVING 145
12. MILK SALES AND CUSTOMER RELATIONS 167
13. THE BIG PICTURE: BELLA FARM 189
    RESOURCE BIBLIOGRAPHY 194

ACKNOWLEDGMENTS

My deepest appreciation goes out to the generations of American family farmers whose shoulders we now are able to stand upon. As the past becomes the future, I am eternally grateful.

This book grew out of my own journey to develop a small-scale raw milk cow dairy for my own family. Along the way, I visited with and learned from many dozens of other dedicated farmers. I learned something from every one of you, and thank you for your willingness to share the art of your craft.

Special acknowledgement goes to Lawrence Holmes, a master Biodynamic farmer whose years of study in Germany brought back a deep knowledge set for how humans and cows can work together on a small diversified farm.

Finally, Joel Salatin, mentor to us all, deserves special mention for his encouragement to take on the task of self-publishing this book. Hearing this advice from the master himself gave me the confidence and motivation I needed to make this dream a reality.

1. THE BEAUTY OF DAIRY FARMING

The dairy cow is the most productive and valuable animal on a small farm. Her ability to convert simple pastures into nutrient rich milk and healthy calves is supreme. In past times, when family farms were the mainstay of American culture, almost all small farms milked a few cows.

As times have changed, and economic and regulatory forces have driven most families into the cities, dairy cows have become resigned to factory work themselves. The days of idyllic green pastures, graced by beautiful dairy cows, have largely slipped away. The agrarian life, enriched by the miraculous biology of the cow, is a life with which many wish to reconnect. It is a good life, an honest and meaningful life, and a life that is still well within reach, even deep in the twenty-first century.

Returning the dairy cow to her native pastures is a homecoming for human culture, and agriculture. The partnership between cow and farmer, a classic pairing of skill and substance, generates miraculous prosperity for farm families and the Earth they walk upon.

Rediscovering a Better Way

The knowledge of how to milk and care for small herds of pastured dairy cows has been long slipping away, reduced to memories and reminiscence. The great majority of dairy farmers nowadays run industrial farms, divorced from the pastoral idyll of days gone by. Fortunately for the future, a few wise farmers refused to be swept along by the currents of modern 'progress'.

The techniques of managing small herds of holistically managed dairy cows are not new inventions; they are timeless traditions. Rooted in nature, with a harmony that generates substance without waste, holistic dairy farming honors the fundamental biological reality of the cow.

Farm families gain in wealth and productivity, by working with a creature of such tremendous fertility and yield. The daily interactions between cow and farmer are a source of physical wellbeing and true security. The partnership between man and cow is wonderful and fulfilling in every possible way.

The Core Principles of Holistic Dairy Farming

Farming in harmony with the needs of the farm family, and the requirements of the natural cow, requires adherence to three key principles. Like legs of a stool, we cannot rest on a stable platform without all three legs standing strong. The three pillars of holistic dairy farming are a pasture diet, milking once a day, and seasonal management of the herd.

Pasture feeding provides for our cows nourishment in the healthiest, and most economical way possible. The cow, with her large rumen and four stomachs, is a wonder at deriving nutrition from the simplest of plant forms. Her entire physical construction is designed to walk through pastures, grazing as she goes, and concentrating the nutrients within her body from simple grasses and forage plants. There is an efficiency in cows grazing that cannot be improved upon, both for her health, and the productivity of our farms.

Milking our cows once a day allows a good life for the dairy farmer. The daily responsibility of milking can be timed in a way that is harmonious to our human lives. In addition, our cows can readily maintain a vigorous and healthy physical condition when we only deplete their milk reserves once a day. Once a day milking is best for both farmer and cow.

Seasonal management honors the biological heritage of the cow, an animal that naturally births in spring, with the onset of lush pasture forage. Her milk production perfectly parallels the growth of fresh grasses, so that her system is nourished in harmony with her milk yield. Calves grow in the momma cow's belly during the long nights of winter, and emerge into a welcoming world of abundance each spring.

There are many details to the proper holistic management of dairy cows. These three principles form the platform upon which all our prosperity can flourish.

The Time is Now

As new generations of aspiring farmers work to develop viable farms of their own, the question of how always rises to the forefront. In our modern era, farming is hugely underrepresented in our society. The way to a successful small family farm is complex and uncertain. Partnering with an animal as productive and emblematic of wealth as the dairy cow is an investment in success.

Having completely ruined the inherent nutritional value of milk, the modern industrial dairy industry has created a sizeable niche for small dairy farms. Human health flourishes in cooperation with the mighty dairy cow, managed naturally, according to her fundamental needs. In a world lacking nutrition, and a society losing its purpose, dairy farming presents a productive way forward.

Rising in tandem with this need for healing food and healthy livelihoods, a renewed interest in the tools of small-scale dairy farming has made proper equipment available to family farms. Consumer demand has created legal opportunities for families to access their food direct from farm producers. Innovations in pasture fencing have made management of cowherds more and more practical on smaller and smaller acreages.

The time is now for a rebirth of the family farm dairy.

Dairy Farming: The Beautiful Way

Beauty is inherent in all healthy life, from a picturesque mountain lake, to the face of a lovely maiden. As we live and work in pursuit of beauty, we are creating natural abundance and wealth in our lives. Beauty is a measure of success for all life.

This book will take the reader through the many considerations necessary for operating a small, holistic dairy. The systems of production and management work together to create a farm that is beautiful: from the milk to the milkman.

This is not a book filled with charming anecdotes or fond reminiscence. This book is written for the future farmer who wishes to incorporate dairy cows into their farm success. Filled with clear directions and detailed information, *Dairy Farming: The Beautiful Way* is designed to give aspiring dairy farmers everything they need to know to begin milking a cowherd of their own, and profitably operating a small dairy.

The text is grouped into four basic sections. First is a discussion of the milk itself, the reward for all our labors. Second is an examination of the cow, and the type of beautiful milking animal that we need for success in holistic dairying. Third is a detailed description of the holistic management techniques that we employ to ensure the optimal health of our cows. Fourth is a consideration of the business side of our dairy, keeping our farms profitable and viable for generations to come.

It is my hope and my belief that this book will make possible the dream of the family farm. By demonstrating the way to manage dairy cows beautifully, I give future farmers access to one of the greatest companions to small farm success. I wish all my readers the highest success in all their future ventures with the venerable dairy cow.

2. REAL MILK

Raw Milk

The fresh, whole milk that comes from the cow is a nearly perfect human food. For millennia, people around the world have esteemed such fresh milk as a priceless asset to human health and wellbeing. Despite the efforts of industrial agribusiness to demonize raw milk, there is a rising groundswell of public demand for real, raw milk.

Raw milk contains many enzymes that work to increase the digestibility of the milk itself. During pasteurization, these delicate enzymes are destroyed. This loss is a significant cause of so-called lactose intolerance. Raw milk contains lactase, an enzyme specifically designed to facilitate digestion of lactose. Raw milk is more digestible to a much larger number of people than its processed counterpart.

The vitamin content of raw milk is superior to that of pasteurized milk. The reason is that heating milk causes the breakdown of many health-providing vitamins. While present in significant quantities in raw milk, Vitamin C, Vitamin A, Vitamin B6 and B12, and Vitamin D are all significantly reduced or destroyed completely during the process of pasteurization.

The active microbes found in raw milk increase the absorption of many minerals, including Iodine, Iron, and Calcium. These, and many other minerals, require specific microbial associations to be properly absorbed by the human digestive system. Raw milk delivers these minerals to our bodies in an absorbable form, increasing the body's mineralization.

The risks of bacterial contamination from raw milk have been greatly exaggerated by the conventional dairy industry, and their proxies working in government. When careful analysis is applied to such claims of illness, it is clear that conventional supermarket foods are far more likely to cause food poisoning than raw milk. The 'health risks' of raw milk are a red herring that do not stand up to any proper investigative scrutiny. Raw milk, especially in the modern context of proper refrigeration and dairy sanitization, is a completely safe food.

When dairy cows are raised in accordance to their natural needs, as advocated in this book, their milk is vastly more nutritious than industrially produced factory milk. Given a proper diet of pasture forage, the composition of cows' milk is drastically different from their conventional feedlot counterparts. Pastured milk contains higher levels of Omega 3 fatty acids, higher levels of Conjugated Linoleic Acids (CLAs), and higher calcium levels. Real milk, from real cows, is superior in every nutritional comparison.

Real, raw milk is a gift to human health and nutrition. It is a complex living food, complete with the enzymes and microbes necessary for its optimal uptake in the human digestive system. The value of real, raw milk readily justifies the work of holistic dairy cow management. Dairy farmers, and their families and customers, enjoy the very best food that nature can provide.

Further documentation on the safety and healthfulness of raw milk is presented in detail by the Weston A. Price Foundation in their publication, *A Campaign for Real Milk*. This compilation of science, investigative journalism, and personal stories is readily available to the public at www.realmilk.com

Colostrum

The first milk that our cow produces after calving is called colostrum. Colostrum is essential for the health of a calf, as it contains the essential microflora that colonize the calf's rumen and intestines.

Colostrum could be considered a super yogurt, perfectly formulated to build the immune system in the calf's digestive tract. Humans too, produce colostrum in the first days after birth, and colostrum is essential to the health of any mammalian baby.

Colostrum is essential for the health of every newborn calf. Without a cup or more of colostrum within the first 24 hours of birth, no calf will ever develop a proper immune system or intestinal flora. When our first calf of the year is born, it is a good precaution to save and freeze a quart of colostrum as a veterinary backup. In case a mother cow died during calving, we would still have a quantity of colostrum on hand to feed her orphan calf. The loss of the cow would certainly be a huge tragedy. But, without a bottle of colostrum on hand, her calf would be lost as well. Saving a small quantity of colostrum is an excellent preventative measure in our veterinary care.

The colostrum content in the milk slowly dilutes over the course of the first week. After a week, there is no longer any colostrum in the milk, and our milk has become normal. During the first days of milking, the colostrum milk will be noticeably thick and yellow in color. It has a strong taste. Its nutritional potency is self-evident. Cow colostrum is a superfood for humans, as well as calves. It is strong in flavor, so a little goes a long ways. Diluted in coffee, or blended into a milkshake, its flavor is completely masked. Consuming a small amount of colostrum with each newly born calf is an excellent boost to the immune system of the farmer and his family.

After ensuring the calf gets a good dose of colostrum when born, saving some colostrum for a future veterinary need, and consuming some colostrum by the farmer and his family, there will still be surplus colostrum produced during the first week of lactation. Colostrum is an excellent nutrient superfood for all the other animals on the farm, and it is recommended to feed any extra to chickens, ducks, dogs, cats, or pigs. Colostrum should never be wasted. It will boost the immune system and health of any animal that consumes it fresh.

The Production Curve

During the first month of lactation, the cow's milk production will steadily increase. She will reach a peak in milk production generally one month after calving. Production should plateau at that maximum amount for a month or so, and then begin a steady decline as her lactation progresses. The production curve is steeper during the increase of the first month, and the decline is much more gradual through the rest of the lactation.

The fertility and breeding cycle of the cow will also influence her milk production. When a cow is in heat, ovulating, her milk production will decline for those couple of days. The drop in milk yield during ovulation is usually just a few pounds of milk. It should be just enough to be noticeable, especially when comparing her milk yields in the daily records. The cow should regain her previous production level swiftly after the conclusion of her estrus.

When the cow is successfully bred, rather than her production rebounding in a few days, it will continue to decline, and these losses in production will not be regained. The nutritional strain of pregnancy will generally cause the cow to drop her daily yield by about fifteen percent. This loss of milk production is typically the first indication of a successful breeding. A significantly larger decline in production indicates a poor quality milk cow, or a cow that is suffering from nutritional stress.

It follows that another obvious factor that can lead to a drop in milk production is a decline in the quality of forage that the cow is eating. Any move to a lower quality pasture will result in rapid feedback when weighing the day's milk. It is very important that we manage our cows grazing to ensure a consistent quality of forage on a daily basis. Losses in production are not generally recoverable until the following milking season.

Losses in milk production will be compounded every day over the course of the remainder of the milking season. For example, when we milk the cows less than precisely 24 hours from their previous milking, the cows will necessarily not be able to produce a full day's worth of milk.

This reduction in yield is typically irreversible, so it is very important that we are punctual with our daily milking time as a means to maintaining our maximum possible production.

To put it in perspective, a decrease of three pounds of milk in a given day during the second month of lactation, multiplied over the next 200 days, will result in a total loss of 600 pounds of milk through the season. Milking will still take the same amount of time, but our total annual yield will be lessened by 70 gallons of milk, from just one mistake in management, with one cow. Good management is critical every day, to ensure that our cows can perform to their maximum potential.

Our cows are very sensitive, and our daily records of milk production serve as an uncompromising assessment of our consistency in the care of our cows.

The Butterfat Curve

An interesting aspect of the cow's production curve is the change in composition of the milk through the lactation. There is a consistent increase in the proportion of butterfat in the milk as the lactation proceeds. This reality is exclusive of variables such as cattle breed or individual cow.

The milk at the beginning of the lactation has the lowest percentage of butterfat. Fortunately the cows' production is peaking in volume at this time. As the lactation proceeds, the proportion of cream in our milk slowly increases. By the end of their lactation, our cows may be giving less than half of the total milk yield that they were at the end of the first month. Fortunately, their butterfat percentage has nearly doubled, such that we may be yielding close to the same total amount of butter each day. Our cows' milk dramatically increases in the proportion of butterfat as the lactation progresses.

3. EQUIPMENT FOR THE PROFESSIONAL DAIRY

Sourcing Dairy Equipment

Small-scale dairy equipment is experiencing a resurgence in North America. There are now excellent domestic sources for all pieces of dairy equipment needed for a small dairy. All milk handling equipment that makes direct contact with raw milk should be made of food grade stainless steel, with smooth welds along all seams. Stainless steel is non-reactive with milk, cleans easily, and with proper care will last forever. Recommended sources include Hamby Dairy Supply in Missouri, Bob White Systems in Vermont, Glengarry Cheesemaking in Ontario, and Hoegger Supply in Georgia. What follows is a piece-by-piece discussion of the equipment needed for operation of a raw milk dairy.

Milking Machine

The Surge bucket milker, introduced in 1922, is arguably the single most useful modern invention for small farms. Until its introduction, hand milking was the only serviceable method for farmers to get fresh milk. For our purposes, the primary objection to hand milking is the cleanliness of milk that it produces. Utilization of a milking machine keeps the milk sealed in a stainless steel vessel during milking, so that no debris or dust can compromise the sanitary condition of the fresh milk. Used properly, the milk machine maintains perfect cleanliness for our raw milk.

Machine milking is also easier on both the cow and the farmer. Hand milking is physically strenuous, and while practical for one or two cows, it becomes a real physical burden to try and hand milk more than a couple of cows. Machine milking works by gentle pulsation of a rubber sleeve that slides comfortably over the cows' teats. The pulsation of the milking machine is actually much more gentle than the action of a human hand squeezing the teat. As such, cows are more comfortable being milked by machine than by hand. The mechanized pulsation is both more gentle, and faster, so that the cow can let her milk down completely in less than half the time with a milking machine. The end result is both more comfortable and more efficient, both for cow and farmer.

Milking machines derive their power from vacuum suction. The sucking action of the vacuum is converted into a pulsating squeezing motion in the rubber inflations that attach to each teat. This pulsing motion alternately squeezes the left side teats, then the right side teats. In this way, milk is squeezed out of two teats with every pulse; unlike hand milking which only works one teat at a time. The vacuum pump does not directly suck the milk out of the cow; it only provides the pulsation in the rubber inflations, which causes milk to flow from the teats. The vacuum pump should be located inside the dairy, with its hose running through the wall into the milk stanchion area.

Surge milking machines are designed with a large stainless steel milk collection vessel that is connected by rubber tubes to the cow's teats. These rubber tubes are housed in stainless steel 'shells'. As the milking machine pulses, milk is released into the rubber tubes, where it drains via gravity into the stainless steel milk vessel.

There are two styles of small milking machines, the can milker and the belly milker. With a can milker, the milk vessel stands on the ground next to the cow during milking. In contrast, the belly milker uses a vessel that hangs from a strap that goes over the cow's back, making no contact with the ground. As the cow lets her milk down, this hanging can fills with milk, transferring the weight from the cow's udder to this hanging can. Filling with milk, the can swings gently back and forth, like a pendulum, from the strap. This gentle swinging creates a subtle 'tug and pull' motion on the cow's teats, which replicates the action of a calf nursing on the cow. This is a real design improvement over the stationary can milker that sets on the ground, as the tugging motion stimulates milk letdown. Another advantage of the belly milker is that if the cow moves slightly within the milking stanchion, the milk can moves with her. A shifting cow can easily tip over the stationary can, resulting in a lot of spilled milk. For these reasons, the Surge belly milker is the preferred style of milking machine.

Surge stopped production of their belly milking systems in 1999, so all belly milkers will need to be purchased used. Complete, refurbished Surge belly milkers can be purchased form Hamby Dairy Supply. This is the recommended way to purchase your milking machine.

The can on the belly milker was made in several different sizes. The largest size can is recommended, which will hold 5 gallons when completely full. When cows are at their peak of annual production, they will often give more than 5 gallons in a single milking. When this is the case, you simply stop milking as the can is nearing full, remove the milker from the cow, and empty the can into your holding tank. The milker is then put back on the cow to complete the milking session. This works fine, and is no problem at all. As a testament to the tremendous abundance of the milk cow, we have yielded an impressive eight gallons of milk from a single milking, from a cow raised on pasture alone.

The rubber parts of the milking machine are still manufactured new, and are readily available from dairy supply companies. There are five rubber pieces in the system: the four teat inflations, and the rubber gasket for the lid. These rubber parts should be purchased new, with a spare set kept on hand in case of a rubber failure, which will inevitably happen at an inopportune time. The teat inflations come in several different sizes, depending on the size of your cow's teats. In general, for good producing dairy cows, you will want the large size inflations.

Besides the stainless steel can, the hanging strap, and the rubber pieces, the only remaining part to the milking machine system is the pulsator. This small device sits on top of the milk can, and transfers the vacuum to the rubber milking inflations. Pulsators can still be purchased new, from any dairy supply company. Your pulsator is set to determine the speed with which the rubber milking inflations squeeze. It can be adjusted, and you will hear the clicking of the alternating pulses increase or decrease in speed depending on the setting. The pulsator sounds much like a metronome, clicking away in perfect time. It should been kept clean, and the slide and rubber pieces inside the pulsator should be lubricated with a little petroleum jelly at the start of each year. Keeping a backup pulsator on hand is definitely recommended, as without a functioning pulsator, you will not have a working milk machine.

To review, the basic parts of your Surge belly milker are the stainless steel hanging milk can, the rubber inflations and gasket, the stainless steel shells and lid, and the pulsator. Purchasing these items all together as a tested and refurbished kit is the best way to ensure that all your parts match properly and function perfectly. Reordering spare parts is then easy, as all the components have identifying numbers on them. Parts for Surge milking systems are generally stocked by Hamby Dairy Supply.

The Dairy

The dairy is the building where we filter, cool, bottle, and process our milk. It can be fairly compact, but it needs to be built to a higher standard of finish than general farm buildings. Our dairy needs to be designed and built like a food service building or restaurant kitchen.

The dairy should be located immediately adjacent to the milking stanchion. A direct path should lead from the farmer's side of the stanchion to the dairy door. This entire area beyond the milk stanchion is forever off-limits to cows, and is the exclusive domain of the farmer.

Fluid milk is very heavy, 8.5 pounds per gallon, and carrying a full bucket milker a far distance will prove to be very cumbersome. A traditional Surge bucket milk machine, when full of milk, will weigh close to 70 pounds. Every foot of distance between the stanchion and the dairy matters when carrying milk. The dairy will be a little room of its own, connected by a tight, exterior grade door, to the stanchion area. It is advisable to have the dairy at the same elevation as the milking area, so that there are no steps that need to be negotiated with a heavy can of milk.

An eight feet by twelve feet building generally will be a comfortable size for the dairy, and can be easily constructed out of standard building materials. Special attention should be made for making the building perfectly rodent proof. The dairy should be well insulated on all sides, including floor and ceiling. This is necessary so that freezing temperatures will not damage the equipment inside. A simple shed roof design, with amply insulated ceiling and metal roofing, will work well.

The interior walls of the dairy should be either gypsum drywall, or smooth wood, either of which will then be painted with a lime wash paint. The *Real Milk Paint Company* makes an excellent, dry powder-based product for painting the interior walls of the dairy. Such calcium lime paint has excellent antimicrobial properties that work to keep the dairy perfectly sanitary. Dirt and debris will have a hard time hiding against the pure white backdrop. Additionally, the white color will create a bright and pleasant working environment in the dairy.

The floor needs to be either poured concrete or grouted tile, ideally installed with an integral floor drain. The floor will need to be mopped regularly, to ensure its sanitary condition. A drain will assist in the washing down process, though it is not absolutely necessary so long as mopping is done carefully. If there is a floor drain, provision should be made for a good plug to prevent entry of rodents or drafts of cold air.

A large window to the south will allow ample sunshine into the dairy. Direct sunlight has a great antiseptic benefit for the dairy, acting as a free ultraviolet sterilizer. Additional windows, arranged to provide good cross-ventilation, are necessary to provide fresh cooling air for the milk-cooling tank. All windows should have screens to prevent entry of flies and insects.

Dairy Utilities

Electricity in the dairy will be needed for operating the milk cooling and milk processing appliances. Three or more outlets, spaced conveniently around the dairy, should be located 2' above floor level. Due to the amount of water and moisture in the dairy, these outlets should have weatherproof covers and GFCI outlets. A good ceiling light with switch near the door is important for nighttime trips to the dairy. A dedicated electrical circuit for the milk vacuum pump is recommended, with its switch outside the dairy next to the milk stanchion, and its outlet inside the dairy where the actual pump will be located.

The water supply to the dairy is best provided with a frost-free spigot that runs up through the floor of the dairy. This way, during the coldest days of winter, the water supply can be cutoff below ground, eliminating any danger of freezing. All water pipes in the dairy can then slope gently downhill to this water source via a flexible hose that can be easily disconnected. Easy draining of all water pipes in the dairy can then be accomplished to prevent any freeze damage to the dairy plumbing.

The water used in the dairy needs to be perfectly sanitary, household quality water. Well water or other treated domestic water are both good choices. Agricultural grade water is not satisfactory for use in the dairy, as it may carry pathogens or other bacteria that could contaminate the milk, and milk handling equipment in the dairy. If in doubt, professionally test your water source before use in the dairy; it needs to be perfectly clean.

Hot water is necessary for washing up milk handling equipment. This hot water will be needed every day of the milking season, so inconsistent systems, such as solar hot water, will not work. The best system for providing hot water in the dairy is a tank less, on-demand propane hot water heater. A large propane tank can be installed outside the dairy, and filled once a year by a propane delivery truck. Electric powered on-demand hot water heaters do not have the ability to heat water hot enough for dairy sanitization needs, and therefore are not recommended. A thankless hot water system will allow you to easily adjust the temperature of the output water, allowing both normal washing and high temperature sanitization. A small and economical model, such as the Takagi TK-Jr, will be perfectly adequate. Installed on a wall inside the dairy, it is protected from the elements. A simple wall vent safely exhausts the combustion gases outdoors.

A 1" black iron propane line will need to run through the wall and into the dairy. This will fuel the water heater, and with a tee and in-line valve, can also provide fuel for a stove. Having a propane burner in the dairy may be desirable for cheesemaking and other milk processing needs. It also serves as a backup hot water heating system in case of emergency.

A 2" drain will be needed for our sink. It is nice to locate the sink in front of the large, south-facing window. This will create a nice workspace, and also will enhance the sanitary benefits of direct sunlight shining on the dish drying area.

Careful consideration of the detailed and specialized utility needs of the dairy is of the upmost importance. The dairy is the most important building on the farm, and should be designed and constructed with this in mind.

Equipment inside the Dairy

Detailed discussion of the individual pieces of equipment for the dairy will follow later in this chapter. For now, it is worth considering the general pieces of equipment needed, their space requirements, and their layout within the dairy building.

A large restaurant-style sink, with three deep tubs and a stainless steel counter, should be located along the south wall of the dairy, to capture maximum direct sunlight. Having hanging hot and cold water fixtures will make washing the large pieces of dairy handling equipment much easier than with standard sink faucets. I recommend using flexible potable water hoses, hanging from the ceiling, with brass ball valves at their ends.

The milk-cooling tank will take up a large area, likely 4' or more in diameter. It should be located conveniently as you enter the dairy, carrying a heavy can of fresh milk. The hanging hoses from the sink should reach comfortably to the cooling tank for daily cleaning. The milk cooling tank will need a power supply, and a nearby fresh air intake, to facilitate the cooling of the milk. This screened vent should be 2'x2', and located 6" above floor height. The vent will need a well-sealing cover to buffer against freezing winter temperatures.

A refrigerator/freezer unit will be desirable for storing processed milk products. Larger units have a higher proportion of usable space, and as such are recommended. Depending on the anticipated uses of the dairy, either refrigerator or freezer space may be more useful. These considerations will be discussed further in the next chapter.

Wire shelves will be good for storing the numerous pieces of small equipment and supplies that every dairy needs. Shallow but wide shelves are best. They can run from floor to ceiling to make maximum use of the floor area. Do not skimp on the amount of shelving you plan; a six foot wide set of wire shelves, stretching from floor to ceiling, is a good guideline.

A cheesemaking vat, if desired, is the final piece of large equipment that would need to be accounted for in the dairy building. Cheesemaking vats are basically identical in size to milk cooling tanks. Depending on the size and design of the cheesemaking vat, 220-volt electrical service may be required. Inclusion of a cheesemaking vat will necessitate a larger dairy building than the 8x12 structure I generally recommend.

In general, cheesemaking on a commercial scale is outside the scope of this book. The regulatory and financial burdens involved in commercial cheesemaking go well beyond the scope of a small-scale raw milk dairy. Nevertheless, if you know that cheesemaking is something you want to incorporate into your dairy plans, now is the time to begin researching and planning for its inclusion in your dairy building. It is always easier and more economical to build things right the first time, than to modify your infrastructure at a later time.

Plan out your equipment layout on paper, to ensure that a correct size building is constructed. Be generous in your space allocations, as moving about with large and heavy cans of milk will require spacious walkways. A well-conceived dairy will be a sacred place on your farm, a temple of worship for the extraordinary food of wholesome milk. Build yours to be the very best.

Vacuum Pump

The vacuum pump provides the 'power' to your milking machine. You will likely want to purchase your vacuum pump at the same time as your milking machine. Complete kits with milking machine and vacuum pump will ensure that you have components that are appropriately sized to one another. The vacuum pumps used in milking systems are standard pieces of light industrial equipment, manufactured by companies like Gast Manufacturing and Thomas Industries. You will want a 'rotary vane compressor/vacuum pump'. This design is quiet, efficient, and durable.

The strength of the vacuum suction can be modified with an adjustment screw located on the top of the unit. Generally, 17 psi seems to be ideal for operating the milking machine, though with your particular pulsator, it may vary slightly. The company you purchase a complete milking machine kit from should be able to help you calibrate the proper vacuum pressure, if needed.

The vacuum pump will be mounted to the top of a vacuum chamber, which is basically a small metal can. If you accidentally milk until the milking can is overflowing, milk will get sucked down the vacuum hose and into this vacuum chamber. Do not worry, this is not a big problem, and it will not damage your equipment in any way. The only problem it creates is that you will now have milk residue in the vacuum hose and chamber. This will need to be cleaned out thoroughly and sanitized with mild dairy bleach. There will be a cover on the front of the vacuum chamber that can be removed for cleaning. The vacuum hose can be disconnected and washed. Make sure to do this promptly so that no rancid milk can collect in your vacuum system.

The vacuum pump is best located inside your dairy, where it can stay clean and dry. It is best to run the vacuum hose through the wall between the dairy and the area with the milking stanchion. Having your power switch for the vacuum pump located next to the milking stanchion is highly recommended, so that you can control the milking machine from where you are working. The vacuum pump also makes some noise when in operation, and the quieter and more peaceful the immediate environment in the milking stanchion, the more contented your cows will be during milking. Contented cows give more milk, a basic fact of dairying that always merits remembering.

Store the vacuum hose hanging from a wall next to the milking stanchion, so that if it is unexpectedly turned on, it will not suck up debris from the barn floor. When it is milking time, the hose is easily taken from a hook on the wall, and connected to the milking machine.

Milking Parlor

In a small dairy, the milking parlor will consist of a single stall. It will be located immediately adjacent to the dairy building. It should be a quiet and calm location, where the cow can be perfectly at peace during milking.

Depending on your climate and weather concerns, the parlor could be a fully enclosed area, or just a roof with open walls. Adequate shelter from the elements, to accommodate comfortable milking during inclement weather, is essential.

A small concrete platform for the milking parlor floor is sanitary and easy to clean. The basic dimensions of your milking platform should be about 42" by 80". This will give you a generous amount of space. You do not want your cow stepping off the sides or back of the concrete pad when you are trying to milk her. You also want ample space to comfortably accommodate the farmer on the milking pad, along with the milk cow.

Cows have a natural tendency to urinate or defecate when nervous, so that should be expected, especially when first training a young cow to the milking routine. Remarkably, over time, the cows will learn to relieve themselves before entering the milking stall. A few calm but firm words with the cow when she does have 'an accident' in the milk stall will quickly correct the undesirable behavior. Nevertheless, a concrete floor that slopes gradually to a drain is essential for maintaining perfect cleanliness.

Given the enormous weight of a mature cow, the concrete should be six inches thick, and heavily reinforced at the time of pouring. The concrete pad can be strengthened by incorporation of any odd scraps of metal pipe, discarded chainsaw bars, and short lengths of barbed wire, all embedded deep within the concrete. These ensure no cracking in the pad will ever compromise its cleanliness.

Be sure to run a section of 2" drainpipe through the rear of the pad, angling downward and terminating outside the barn, to act as a simple drain. This drain can serve as a simple gray water source for watering shelterbelt trees in the cow and calf yards.

When pouring the concrete slab, make sure to lightly score the surface, to ensure that it is not at all slick when wet. Additionally, scrape a gentle groove around the edges, which drains back into the drainpipe, so that when washing down the pad, the surrounding floor stays relatively clean and dry.

The Milking Stanchion

The milking stanchion needs to be built very solidly. First time milk cows may become uneasy and decided to 'leave the building' at an inopportune time. Given the strength of a full-size milk cow, good construction and materials are imperative. If you have good welding skills, metal fabrication from cut and welded steel pipe makes a superior milk stall. Otherwise, 4x4 lumber, held together with large lag screws or bolts, will work well. The posts that you put into the ground to anchor the stanchion should be solidly anchored, 6" round posts. These should set into the ground just outside the perimeter of the concrete pad you have poured.

Looking at the four sides of the milking stall, you can think of them as the front (where the cow's head is), the back, the farmer's side, and the cow's side.

Starting with the cow side, this is where she will enter the milk stall. You want an out-swinging metal gate at the rear of the cow side. The cow will step up onto the platform, and move forward to the front of the platform. You can then close the gate behind her. The stall must be narrow enough that once inside, the cow cannot turn around. She will back out from where she came when milking is finished. This will be no problem for the cow. You want to make her entry easy; she will exit willing, even going backwards.

The farmer's side of the milking stall is conveniently made from an 8' tubular steel cattle gate. It will be mounted to the post at the front of the milking platform. The rear of the gate can then be secured to the gate on the cow's side, by running a simple chain behind the cow once she is in the stanchion. The chain can be secured at whatever length is appropriate for the size of a given cow. This chain also prevents the cow from backing up once she is in place and ready for milking. The slight contact on the cow's sides and rump makes her feel secure, and reduces any tendency to want to move. Cows are comforted by gentle pressure on their sides and rump.

The design of the two gates serves three functions. First, it creates a bright and well-lit enclosure, which makes loading the cow easier. As the cow enters the milk stall, she is looking at a gate, which except for the metal bars is mostly open space. This is good, because cows are inclined to move forward towards light. The cow will step willing into the stall, and then be guided by the metal bars, into position.

Second, the gate design leaves a 'window' for access to the cow's udder. Leave the entire frame of the gate intact, and cut out two bars from the bottom of the rear section of the gate, so that there is a simple opening directly in front of the cow's udder. This window should be about 2 feet high by 3 feet wide. This way, the farmer can be 'outside' the milk stanchion, safe from being crushed by the cow, while still having good access to her udder, and still standing on the concrete pad. The farmer comfortably reaches through this window for udder cleaning and attaching the milking machine.

Third, the gate can swing open or closed to adjust the width of the milking stanchion for each individual cow. When welcoming the cow into the milk stanchion, leave the gate set in a wide position, so that the stall seems large and inviting. Once the cow is in the stall, gently close the gate snug against her side. We want to keep the stall narrow enough that a cow cannot turn around once inside. When it is time for her to leave the stall, the chain is removed from behind her rump, and she is free to back out at her own pace.

The rear of the stall should just be made of solid material. There is nothing that happens here, so it can be simple. An existing barn wall works great for the rear of the milking stall. Placing a panel of smooth painted plywood or other easily cleanable material here is a good idea, since this is the manure end of the cow, and inevitably, that will happen. It should be noted that the rear of the stall also could be used as the entry location for the cow. If this were the case, then the 'cow side' would just be a solid wall. The two walls, the cow side and the rear, would simply be reversed in function.

The front of the milking stall should have an opening that the cow can comfortable put her head through. A traditional 'head catch' is not necessary. In fact, it is disrespectful to the cow and counterproductive for her docility. Our milk cows willingly give us their milk, and physically restraining them by their head makes them afraid. Cows like to be squeezed gently from their sides, like in a traditional veterinary 'squeeze chute', but they do not like to be restrained by their neck. We need to work with the cow's instincts to keep her calm and contented. I think that head catches make cows anxious, and that works against their milk letdown, which is a highly emotional and maternal process for the cow.

The opening for the cow to place her head through should be roughly 2' by 2'. This is large enough to accommodate mature cows with full size horns. The opening allows the cow freedom of movement with her head, while making it impossible to swing her horns around, or rear upward with her head. These movements from a startled cow could be dangerous for the farmer. Especially when a cow first freshens, her udder can be very tender and painful. We do not want her instinctive reflex to this pain to expose us to any direct danger of being horned.

A trough immediately in front of the head opening enables the cow can enjoy a lick of kelp and salt while milking. Do not feed cows in the milk stanchion; it is an unnecessary distraction for the cow, and an unwanted expense for the farmer. Once the cow comes to understand the milking process, she will willingly enter the stall and stand calmly for milking. Her reward is the natural endorphins that are released when she lets down her milk. Kelp and salt are offered in the milk stall to ensure daily access to these highly mineralizing health supplements, the dairy farmer's best form of health insurance for his dairy cows.

The milking stanchion doubles as a useful location for any veterinary treatments that become necessary. In a holistically managed dairy, this will almost never occur. As such, the cost of a traditional cattle squeeze chute for veterinary treatments cannot be justified for the holistically managed small dairy herd. Nevertheless, emergencies are best planned for ahead of time, and knowing that the milk stall can be used to administer a veterinary treatment to an ailing cow is a nice insurance to have.

A stool for the farmer should be located just to the side of the 'udder window'. It should fit nicely on the concrete platform, as the platform size recommended above is larger than needed for just the cow to stand upon. This stool on the platform gives the farmer a comfortable seat for cleaning the udder. It also places the farmer in the natural position of the calf, which encourages the momma cow to see us as her baby. In this way the cow becomes more eager to give us her milk.

Consider the cow's 'view' from the milk stanchion. Cows are very sensitive to moving objects, so the environment in front of the milk stall should be very calm and quiet. Remember that cows see shape and movement far more than color or detail. The book by Temple Grandin, *Humane Livestock Handling,* is particularly illuminating on this subject. We want to create an environment that is peaceful and comforting for the cow, much like the environment she would seek out in nature for nursing her calf.

Cooling Tank

Rapid cooling of your fresh, raw milk is one of the most important steps for ensuring top quality milk. The quicker that milk is chilled to 34 degrees F, the better its nutrients are preserved, the sweeter its flavor will be, and the more sanitary the product. All fluid milk must be completely chilled to this storage temperature within 2 hours of milking.

In small quantities, milk can be cooled in 2 or 3-gallon stainless steel cans, submerged in an ice water bath. It should be cautioned that this method requires large amounts of ice to work properly. Milk cans full of warm milk need to be positioned to have 6 inches or more of ice water surrounding all sides of the can. This system works well for a family cow, or at the beginning or end of a milk season when only one cow is milking. It becomes nearly impossible to maintain the necessary cooling schedule with multiple cows. Above all, monitor the temperature of your milk in the cans, and be sure that you are cooling your milk completely, within the recommended timeframe.

For a small commercial raw milk dairy, a proper milk cooling bulk tank is an investment in quality. It will be the single most expensive item for a startup dairy. Bulk cooling tanks are now manufactured in the United States, and sold by Bob White Systems of Vermont. This is a huge improvement on just a few years ago, when small-scale milk cooling tanks had to be imported from Holland at considerable expense.

The recommended size for a bulk-cooling tank is 50 or 60 gallons, depending on your herd size. A good guideline is to figure on having 9 gallons of storage capacity per milk cow. Generally, this will allow you to combine two days of milk before bottling or processing, which will prove to be useful time after time. Given the minimal difference in price for different size tanks, it is recommended to err on the larger size for your cooling tank.

A professional bulk milk-cooling tank will cool the milk via a double-jacketed tank design, with refrigeration coils running around and beneath the entire milk holding tank. An agitator paddle slowly circulates the milk, so that warm spots are eliminated during the cooling process. A thermostat will automatically turn on the refrigeration loop when the milk rises above a set temperature point. The agitator paddle will cycle on every twenty minutes to prevent the cream from separating and rising to the top. In this way, your milk is chilled to perfection, and then stored sanitarily until you are ready to bottle or process.

A large, full-port ball valve at the bottom of the tank can be used for bottling. A sturdy valve handle will make controlling milk flow easy. A downspout from this valve will direct your milk into milk jugs, glass jars, or large transport cans. The ball valve will disassemble easily for sanitization. With a proper milk tank, there is no need for specialized equipment for milk bottling. The valve on your cooling tank is also used when washing out the tank between uses. You can spray out the inside of the tank easily with the hanging hot and cold water hoses that you have installed above the sink in the dairy. This wash water can be drained into a bucket, and poured down the drain of your dairy sink for disposal.

Separating Cream

The ability to easily separate the cream from your fresh milk is one of the best things about cow's milk. As it always has, the cream still rises to the top of natural, raw, whole milk. On a small scale, the housewife can simply place a half-gallon jar of milk in the fridge overnight, and then spoon off a half-cup in the morning for the day's coffee. It works wonderfully, and is a real treasure.

On a larger scale, the ability to separate large amounts of cream for butter making is very important. Specialized cream separator machines do this job very effectively, but are expensive and complicated. A small cream separator contains more individual parts to be washed than the entire milking machine. Assuming that you would want to be able to separate a large volume of milk, such as a day's entire milking, it would require a small scale commercial cream separator, which would cost several thousand dollars. There is a better way to butter!

Since cream naturally rises to the top, we can use our bulk tank as a large, gravity cream separator. In the evening, once the milk is fully chilled, simply turn off your milk tank. This will interrupt the regular cycling of the agitator paddle, so the cream will steadily rise to the top. By morning, there will be a thick layer of cream at the top of your bulk tank. Yield will range from 15% cream per volume of milk at the start of the milk season, to up to 35% cream at the end of the milk season, as the percentage of butterfat slowly increases through the milking season. This percentage will also vary by breed and individual cow, with Jersey cows certainly giving a higher percentage of cream than Holsteins.

With the cream naturally separated to the top of the milk tank, now you can slowly drain the skim milk off of the bottom of the tank. Opening the valve on the tank slightly, the skim milk slowly drains without agitating the stratification enough to disturb the perfect layer of cream that is sitting on top. You will drain off bucket after bucket of skim milk, which can be used for cheesemaking or to feed to your chickens. Watching carefully, you will suddenly notice that the milk that is draining has changed in color and thickness. This is your raw cream! Close the valve, count your blessings, and grab a new container to fill with your precious raw cream. What remains in the tank will be raw cream, good to the last drop.

Using your bulk-cooling tank as a low-tech cream separator is an excellent way to get the best value out of your most expensive piece of dairying equipment. The cream you yield is the most precious and valuable product of the dairy. It can be consumed directly as raw cream over fresh berries, processed into farm fresh cream cheese, or churned into the gold of dairy farming, raw butter.

4. THE DETAILS OF MILKING

A consistent routine is the most important thing in the daily process of milking cows. A well planned, carefully organized, and patiently executed routine allows the cows to be more calm, more content, and as a result, to give more milk. Cows are intelligent creatures, who learn their milking routine, and benefit greatly from daily consistency. As a dairy farmer, take pride in establishing a good rhythm to the daily milking routine. Remember to always consider how your routine is for the cows, and to honor their consistency in production, by maintaining a consistent milking routine.

Our daily routine involves herding the cows in from the field and into their individual stalls in the barn. We then assemble the milking machine, bring our first cow into the milking stanchion, brush her, clean her teats, massage her udder, and hand strip each quarter of her udder. Now we attach the milking machine and milk our cow. When finished, we weigh the milk and pour it through the filter and into our cooling tank. Finally, we hand strip the milk in each quarter, return the cow to her stall, and when finished with the entire herd, let them back out to pasture.

The first and most important aspect of our milking routine is the timing of our daily milking. Dairy cows need a fixed time to be milked each day. Assuming a once-a-day milking routine is followed, the farmer can choose anytime he or she prefers. But once a time has been chosen, milking needs to occur at the same time precisely, every day. We can gradually shift the time of milking by twenty minutes each day, without ill effect. This is particularly useful in the fall as the days are getting shorter and the mornings colder. We can slowly shift our milking from 8am to 11am over the course of several weeks. But we absolutely cannot be erratic in the timing of our milking, or our cow's production will suffer irreversibly.

Herding Cows to the Barn

The daily milking routine begins when you walk out into the pasture to gather the cows. They will immediately lift their heads, and recognize your presence. From this moment forward, all efforts should be made to handle the cows in a way that keeps them perfectly calm. Chasing cows through the pasture, or herding them with a barking dog, are both wholly unacceptable. We approach our cows gently, as animals to be trained, so that they learn the expectations of our daily routine. Patience is the tool that we use to teach our cows.

Cows are best moved as a herd, not as a collection of individuals. Being a herd animal, cows feel comfort in a group, and distress when isolated. As such, our first step in moving the herd is to group all of our cows together in the field. One by one, we move the cows towards each other, creating a tight herd. This will make the cows emotionally comfortable, so that they are less flighty when being herded. Working with the cow's natural instincts achieves easy results for the farmer.

Herding cows is a simple process, with a few basic rules. Cows like to move forward, and they turn from their front shoulders. Knowing these preferences of cow physiology, we can work with the cows' natural instincts to direct their movement. When we move towards the cow's rump, it is like pressing on a gas pedal, accelerating them forwards. We call this applying pressure. When we back away, the cow slows down, and ultimately comes to a stop. If we want the cow to turn, we move towards their front shoulder. They will turn away from us, so we can control their direction. Combining these two variables, of distance from rump and distance from shoulder, we can control speed and direction.

The positioning of our body in relation to the cow is our primary means of directing her movement. When this is not enough motivation for a lazy or obstinate cow, we have additional strategies. The use of our body position, followed by our outstretched hands, then an extended stick, and finally a tap with the stick, gives us four escalating techniques for training our cows' movement.

When you need to 'encourage' your cows a bit more to turn or move you can lift your hands to apply extra pressure. One step further, a long stick can act as an extension of our arms. Under the best of circumstances, we never even show the cows that we are carrying a stick. We never want to be waving a stick around like a maniac; this would just distress our cows. If needed, we hold our stick out to the side, making our arms appear much larger, for emphasis. Finally, a particularly stubborn cow can be encouraged forward with a gentle tap of the stick on top of her rump. We use the stick as a tool of last recourse, for training not for discipline. Never use a heavy tool when a lighter one will suffice. Patience is always more persuasive than force.

When moving the herd, our first step is to get the cows heads pointed in the correct direction, since cows naturally want to move forward. We apply pressure from behind, calmly walking back and forth behind the cows, slowly directing our cows towards the milking barn. If an individual begins to turn away from the group, we sweep around to the side, steering the wayward cow back into the herd. Moving forward, we want to maintain just enough pressure to keep the herd moving slowly and steadily. We do not want to be so close to the herd that it agitates them and they begin to move quickly. Cows only move fast when distressed, and we want our cows calm and comfortable in anticipation of milking.

Cows are as easily trained as horses, and quickly learn that they are being herded to the barn for milking. The process of animal training develops through a series of stages, in a systematic way. First, the cows anticipate our presence in their pasture at ten in the morning. Second, the cows act upon that anticipation and begin naturally moving toward the barn when they see us entering their pasture at that time. Finally, the cows will remember the entire process to the degree that they will walk themselves to the barn at ten in the morning, and be waiting for us, right on time for milking. In all our interactions with the cows, we must remember the saying, "Whether or not we are teaching, the cows are always learning."

Cow Stalls

In a small dairy herd, having individual cow stalls is an affordable luxury that will make milking time more efficient and cattle care easier. 10 foot by 10 foot box stalls, as are typically used for horses, are ideal. The cows will quickly learn to recognize their personal stall, and will come into the barn for milking, walk straight into their stall, and lay down waiting their turn to be milked. It is recommended to have a single structure that contains the cow stalls and the milking parlor, as this will be most efficient both for the construction and the management of the cow dairy.

The walkway leading into the barn and to each cow's individual stall should be ample in width. Five feet wide ensures comfortable and stress-free movement of the cows into their stalls at milking time.

Boards for the sides of the cow stalls are recommended because they provide the cow privacy when laying down. 2x8 lumber is an ideal choice. The milking barn must maintain a quiet and peaceful atmosphere. Having stall walls that keep each cow in her own private environment will keep all the animals much calmer. There will be no external sights to distract the cow. Calm and contented cows make better milkers, and produce more milk.

Sawdust is the preferred material for the floor in the cow stalls. A thick bed of sawdust, a foot deep at minimum, will readily absorb any manure or urine that the cow deposits. The sawdust should be sprayed each spring with Biodynamic preparations or compost tea, to inoculate beneficial bacterial that will outcompete the growth of other, potentially unhealthy bacteria in the dairy barn. Allowing chickens access to the cow stalls, during times of day when milking is not occurring, will help to aerate and mix the sawdust compost, further increasing the health of the bacterial environment in the cow's bedding.

Sawdust is much easier to shovel by hand than woodchips. Any large manure patties can be shoveled into a wheelbarrow and brought outside to add to the compost pile. Small amounts of sawdust can be added annually to refresh the surface. The amount of work needed to keep the cows' stalls clean and sanitary is quite minimal.

In the event of a calf being born during particularly bad spring weather, the sawdust bedding in the stall should receive a fresh topdressing. This dry sawdust provides a warm and clean environment for calving to occur. Generally, this will be unnecessary, as outdoor calving is strongly preferred, but having good stalls where the cows are already comfortable and habituated, will make the farmer's job that much easier in difficult veterinary circumstances.

Large openings along the tops of the walls will be desirable for both ventilation and light. By locating the openings six to eight feet above the ground, ventilation will occur without cold drafts at ground level. These large openings will also allow in good ambient light. Good location of these vents on the ends of the building will make fans completely unnecessary. We allow nature to provide clean air through good building design. There never is a need to warm the barn air above ambient temperature, so it is unnecessary, and even undesirable, to have a building that seals tightly.

Having a single water spigot in the barn, with a rubber hose and a few moveable rubber watering buckets, will adequately accommodate the rare need to give a cow water in her stall. Similarly, hay feeders are not needed on a permanent basis. Having several rubber tubs on hand, that can be set out if needed for feeding, will be more than adequate. These rubber buckets and tubs are preferred because they will not get cracked and broken like plastic or metal ones would.

Entering their Stalls

Before we head out into the pasture to gather our cows, we will have checked the barn to make sure that all gates are in their proper position. The main entry gate to the barn will be wide open. The gate to the milking stanchion should be closed. Each gate to the cows' individual stalls should be open. Any odd pieces of clutter or tools should be put away, leaving nothing out of place that could confuse the cows. We want the barn to look exactly the same, every time the cows come in for milking. Consistency creates comfort for our cows.

When we first introduce our cows to the barn, it is best to let them choose their own stalls. The traditional ten-by-ten box stall is small enough that the cows will view each stall as the territory of an individual. A more 'boss' cow will push out a 'lower' cow that tries to enter her same stall. This process will usually take a few attempts to sort itself out; that is fine, use patience as your tool. The less we interfere, the more quickly our cows will figure out ownership of their own individual stalls. Given a chance, they will establish and maintain their own order in all things.

Walking in from the pasture, our cows will develop a natural order, the same cows leading and following one another into the barn. They will enter the barn in a single file line, nose to butt, moving in the self-established order that they find most comfortable. As the cows enter the barn, we should back away from them, not pressuring them too quickly. We let the cows calmly sort themselves into their stalls. Once the cows are comfortable in their individual stalls, we enter the barn and close the main gate behind us.

Now is a good time to reflect on the fact that our cows are large, powerful animals. Out in the pasture, a startled animal has the opportunity to flee a stressful or confusing situation. Inside the barn, our cows cannot simply move away from something they find uncomfortable. Handling cows in the confined environment of the barn can be dangerous, and should always be treated with respect and caution. If a cow is not moving where we want her to, we have to always remember to keep 'safety first', and never place ourselves in a position where we could be crushed. Using our long stick to gently redirect a wayward cow is much safer than entering a confined stall with a large and confused cow. Patience, as always, is our best tool for good results. If all else fails, and the cows will not move where we want them, walk away. Give it ten minutes, and come back and try again. Persistent patience in the proper position is our motto; it will win every time.

Once situated in their stalls, the cows should quickly lie down, and contentedly resume chewing their cud. We quietly close the gates to their stalls, and acknowledge each cow by name. Cows learn their names, they know who they are, and appreciate the respect of our regular greeting. While the cows relax from their walk in from the pasture, the farmer heads into the dairy to get the milking equipment ready.

Assembling the Milking Machine

The components of our milking machine should be waiting for us, neatly organized on the drying rack in the dairy. All parts should be completely dry from the previous day's cleaning. Pull the four rubber inflations into their stainless steel shells. Insert the rubber gasket into the lid of the bucket milker. Place the lid on the milking can, and attach the pulsator to the lid. Connect the milking inflations to the lid and the pulsator, being mindful of any twists in the rubber hoses that could become kinked when the machine is attached to the cow. A little practice and experimentation is worthwhile to figure out the best orientation of the rubber hoses. An ounce of prevention is worth a pound of cure.

Our milking machine is now ready. Bring the milking machine out to the milking stanchion, and set it on the clean concrete floor of the milking stall. Attach the vacuum hose to the pulsator, and turn on the vacuum pump, listening for the rhythmic clicking of the pulsator to ensure that everything is working properly. Once proper operation of all four inflations has been confirmed, turn it off. This process of assembling and checking the milking machine should take about a minute.

Into the Milk Stanchion

In any herd of cows, there will always be one cow that is the natural leader of the herd. She could be described as your best cow, your boss cow, or your top cow. It is best to milk her first. The other cows will watch her, and tend to imitate her behavior, since she is the natural leader of the herd. Milking the head cow first establishes a good pattern for the others to follow.

Open the gate to the first cow's stall, and verbally tell the cow, 'now it's time for milking, get in the milker.' We will calmly repeat this refrain several times, using the cows name as we talk to her. Occasionally the cow may be feeling lazy, and a light tap on her rump with a stick will encourage her to stand up and head to the milking stanchion.

The cow will generally urinate when she first stands up, as to not later have to urinate when in the milking stanchion. We verbally thank her for this act of consideration. If she neglects to urinate before leaving her stall, we remind her to go pee before heading to the milker. In time, urinating before entering the milking stanchion will become second nature, a trained habit that benefits the farmer. It is amazing the amount of verbal understanding that our dairy cows develop with time and conscious training.

We allow the cow to slowly make her way to the milking stanchion. A well-trained cow will come to understand this entire routine, and will move purposefully to the milking stanchion. She may pause as she steps onto the concrete milking platform. Remember that cows have generally poor eyesight, and may need a moment to assess their situation and know exactly where to move to get into position. We stand behind the cow, so that she is not inclined to back up in a confused moment. We apply patience to the situation, but also may need to gently tap the cow on her rump to encourage her forward. Once she steps up fully onto the concrete milking platform, we close the gate behind her and walk around to the farmer's side of the milking stanchion.

The cow should now step fully forward, placing her head through the large opening at the front of the milking stanchion. She may take a lick of kelp at this time, as it is like a health food candy for cows. We slide the farmer's side gate snug up against the cow's side. The gate's closure chain is pulled across behind the cow's rump, and secured to the entry gate. The cow is now comfortably secured in the milking stanchion, in a way that allows her to shift her weight side to side, but not walk forward or back. She is secure, both physically and emotionally, and we are ready to begin milking.

Brushing and Cleaning

Gently talking to the cow, we brush her with long strokes that flow from front to back, in a slow and methodical manner. Brushing the cow stimulates her endocrine system, which is deeply connected to her milk production. Brushing invigorates her circulation, and awakens her senses. Brushing our cows is about more than just keeping them clean and healthy. Studies from Germany have shown that brushing cows results in them giving more milk. The brushing is an act of kindness on our part, and we are repaid for our kindness, by their generosity in letting down more milk.

If the cow has been in muddy conditions, a metal currycomb will be needed to dislodge any dried mud clumps on her coat. We want to keep our cows clean, for their health and for our milk quality. After any dried clumps of mud are removed, we use a standard bristle brush. Starting at the front of the cow, we brush downward from the spine, brushing in the direction of the hair growth. We brush down her shoulders and ribs, down her sides and rump. Cows seem to particularly enjoy being brushed on their chest, down their breastbone, right where their heart is located. The final step in brushing is to brush down the cow's spine, from front to back, in long, slow, sweeping strokes. A properly brushed cow will appear sleek and shiny, glossy and radiant, regal in her wellbeing.

We now brush the cow's udder, remembering that the cow's teats are sensitive, quite unlike the thick hide that covers the rest of her body. The udder needs to be brushed perfectly clean, free from even the tiniest specks of mud or debris. Brushing cleans the udder, while also stimulating the udder to encourage milk letdown. Even the teats themselves get brushed gently with the bristle brush, exfoliating any dead skin. No matter how dirty, we never use water or a wet cloth to clean the udder. Brushing is the mechanism for cleaning the teats and udder in preparation for milking.

The teats are now sanitized with a disposable teat wipe. Teat wipes are purchased in gallon buckets, and we use one teat wipe per cow at each milking. The teat wipes contain sanitizing agents like alcohol and essential oils. They also contain soothing and moisturizing compounds like lanolin that keep the teat tissues supple and healthy. Good care of the teats is critical, as the skin on the teats is very sensitive. If the teats dry out, the skin can crack, resulting in a very painful situation for the cow which will make milking difficult. Therefore it is critical that we use specially formulated teat wipes that both clean and nourish the cow's teats. Teat wipes are easily purchased in both small and large quantities from Hamby Dairy Supply in Missouri, or other dairy suppliers. If our cows' teats become chapped, application of a moisturizing salve after milking is an essential practice in holistic cow care.

Massaging the Udder

Every time before milking, we need to thoroughly and strongly massage the udder bag. Using the rounded form of our fist, we push and knead the udder. This will be a physical and tiring process for the farmer. We are not 'punching the bag', but we are being quite strong with our massage. Our daily udder massage performs the same biological function as the calf butting the udder while nursing. Massaging the udder stimulates the cow's hormones to let down her milk. Like a well-worked muscle, the udder can be sore, and massage may be painful, but it ultimately is healthy for the swollen tissues. A good udder massage is critical for good milk letdown, and long-term udder health.

When our cows first freshen, their udders become tightly engorged with milk. The tissues in the udder are almost always swollen and painful. To help soothe this inflammation, there are special udder lotions that we can apply directly to the udder bag. The commercially available product that works best is called Dynamint. It is a menthol lotion, which cools the skin and reduces inflammation. Dynamint is sold in quart and gallon sizes, and is expensive. Generally we only need to use the Dynamint lotion for the first week or two after calving. After this time the udder will have softened and lost its excessive engorgement. The cooling essential oils in the Dynamint make the cow much more comfortable, and further stimulate her let down of milk. The use of Dynamint in the week after calving is one of the best preventatives for mastitis. It also makes milking easier and higher yield, and helps our cows to be more comfortable with their engorged udders.

Any hard spots in the udder need to be thoroughly massaged with this medicated lotion. Hard spots may even feel noticeably warm to the touch. This is an indication that there are blockages in this part of the udder, which can lead to mastitis. Massage is the best way to break up these blockages, keeping the udder supple and healthy. The cow may be uncomfortable when we massage these spots, as these blockages in the udder are generally inflamed and painful. We need to be gentle yet firm, realizing that the long-term health of the udder depends upon our care in resolving these blockages. Thoroughly massaging the udder is one of our most important responsibilities in the daily milking routine.

The Strip Cup and Test Kit

Before we attach the milking machine, we will hand milk 8 squirts of milk out of each teat. Hand milking is a bit of an enigma until you have practiced a bit. You are not pulling on the teats, or randomly squeezing the teats back and forth. The best way to understand hand milking is to think of the teat as a little sausage, with an opening into the udder on one end, and the opening of the teat on the other. We want to first pinch off the top of the sausage so that no milk can push upwards into the udder. For this we use our thumb and forefinger. With the top of the teat sealed off, we now squeeze the milk downward, out of the teat opening with our other three fingers. Milk will squirt out in a strong stream. We now release the teat entirely, let it refill with milk from the udder for a second, and repeat. Pinch off the top of the teat, and then squeeze out the milk. We need to wait long enough for the teat to completely refill with milk, which will vary from cow to cow. Go slow at first, being perfect in your motions. Speed will come with proficiency.

These eight squirts of milk per quarter are milked into a 'strip cup', which is just a simple plastic cup with a screen top. There are several reasons for using the strip cup. First, it catches the small plug of dried milk that forms at the tip of the teat. This natural seal is formed to keep anything from entering the teat canal. Second, it clears the milk that has been stagnant inside the cow's teats. We want the milk that is healthily stored in the cow's udder, not the last bit that is in the teats themselves. Third, it gives us a small sample of milk that we can examine to make sure that there are no mastitis issues developing in the cow's udder.

As the milk passes through the screen in the strip cup, any abnormalities are easily observed. Healthy milk from a healthy udder should be perfectly clean and uniform. We are watching for any clumps, stringy bits, or chunks that come out of the cow's teats. We observe on a teat-by-teat basis, ensuring that all four quarters are passing perfectly healthy milk before we attach the milking machine.

If any abnormalities are observed, it is necessary to further evaluate the milk. A simple testing kit, called a California Mastitis Test Kit, is an instant and inexpensive way to check for elevated bacteria levels in the milk. This is a primary indicator of clinical mastitis.

The test kit has a paddle with four separate cups, one for each teat. A small amount of testing solution is placed in each of the cups. One by one, we squirt a little milk from each quarter into the respective cup on the testing paddle. Remarkably, if there are elevated bacteria levels in the milk, indicating mastitis, the milk will change color. If the milk is perfectly healthy, the milk will look normal. Thus, the California Mastitis Test Kit lets us know immediately if we have a problem with our cow's udder health and resultant milk quality.

The four quarters of the cow's udder are completely separate from one another, so mastitis in one quarter does not affect the milk quality in the other quarters. If one quarter reacts during the mastitis test, we determine which quarter has the issues, and do not milk that quarter with the milking machine. We will milk the three good quarters with the milking machine, as per usual. When finished, we then hand milk the mastitic quarter into a separate container. Most importantly, we need to massage the problem quarter aggressively with our free hand while milking. Be extremely thorough to make sure that every last drop of milk is emptied out of the problem quarter, as this will help the cow to clear out the mastitis infection. This milk is fine to feed to farm animals, but should not be consumed by humans.

Mastitis is a common affliction in commercial dairy herds; however, it should not be considered acceptable in a small raw milk dairy herd. A cow experiencing a short bout of mastitis, especially just after calving when the udder is massively engorged, simply needs a little extra care and attention. In general, the tolerance that a raw milk farmer has for mastitis is much smaller than that of a conventional dairy farmer. Any cow that experiences prolonged periods of mastitis, or mastitis later in the lactation, or mastitis in multiple quarters, should be culled from the herd. Mastitis is indicative of an unhealthy udder, and we need cows with perfectly healthy udders to produce top quality raw milk. Good dairy cows will never experience mastitis, and hopefully this will be your experience. The section later in the book on selecting dairy cows will help to ensure that you build your herd from animals that are robust in health and never experience health problems like mastitis.

Showtime!

The cow is standing placidly in the milking stanchion, she has been brushed and cleaned, we have cleared each of her teats, and the milking machine is setting on the concrete ready to begin. We loop the bucket milker strap over the cow's back, attaching it at the proper height under her belly. The strap should be located far forward on the cow's spine, just behind the cow's shoulder blades. The height of the strap should be as low as possible without the belly milker can dragging on the ground. This arrangement of the strap forward and the can hanging low, accentuates the tug-and-pull motion on the cow's udder. The cups of the milking machine should be pulling down and forward on the cow's teats, imitating the action of a nursing calf. If the cups do not stay attached to the teats, and are pulling off, then you need to raise the height of the can or slide the strap back a little. Each cow is shaped differently, so experimentation will be needed to find the perfect fit. Remember that the goal is a low can and a forward strap. With this objective in mind, then find the right balance for each individual cow.

With the strap in place, turn on the vacuum pump, and the milking machine begins clicking with life. Hook the belly milker can over the wire on the strap, and attach the inflations one by one to the teats. It is important that the vacuum tubes that connect the inflations to the pulsator do not get pinched, or they will not work. The inflations should be attached quickly, since their suction on the teats is what keeps the machine from pulling off. Once all four inflations are in place, we can feel the vacuum tubes to ensure no pinching. Then feel the base of each inflation tube, noting the warm milk draining into the milk can. Once everything is in its proper place, slowly step back, and let the cow do her work.

The pulsator rapidly clicks along, the cow stands chewing her cud, and we hear the sound of milk filling the metal belly milker can. In a perfect world, it will take about 15 minutes to milk out a cow completely. Nothing is required of the farmer during this time. We maintain a quiet and peaceful environment in the dairy, and observe with awe the remarkable yield of the dairy cow.

As the cow's udder empties, one by one, the teats will dry up, and that inflation will begin making a slurping or sucking sound. This lets us know that a given quarter is emptied. We now pinch the base of that vacuum tube to stop the pulsation, gently remove the inflation, and let it hang down naturally to the side of the belly-milking can. Once the inflation drops down into that position, its suction is cutoff, and we do not need to worry about it. As the inflations are removed one by one, the total amount of suction on the cow's teats reduces, and we may need to hold the can so that the last inflations are not pulled off prematurely. At last, all four quarters will be emptied, the inflations will all be removed, and the vacuum pump can be turned off.

We now lift the full milk can out from under the cow. The empty belly milker weighs 23 pounds, so now full with up to 5 gallons of milk, it will be quite heavy. We immediately carry it into the dairy, and set it down on a digital scale on the counter. We weigh our milk for each cow every day, and record our yields in a notebook. These records will be useful for business planning and cow comparison over the years. Once weighed, the milking can lid is removed, and we carefully pour our milk through the milk filter and into the bulk cooling tank. We can feel proud of our skill as a dairy farmer, producing such a volume of nourishing raw milk.

The final step in milking is to hand strip the remaining milk out of each quarter. This is essential for maintaining perfect udder health. One quarter at a time, we strongly massage the udder bag while hand milking the teat. This milk can be collected into the test cup. Draining every last drop of milk prevents any milk from stagnating in the udder, where it could foster unwanted bacterial growth. The milk we yield from hand stripping the cows is best fed to farm animals.

In a good dairy cow, there should be almost no milk held up in the quarter after removing the milking machine. But for a young cow, or a cow with a particularly engorged udder, there can be quite a bit of milk that will come out with a vigorous massage and hand milking. At this point, with the udder mostly empty, any hard spots indicating blockages will be more apparent than during the pre-milking massage. Repeatedly massaging these hard spots, while stripping from the udder, will break up these blockages. This is a critical part of dairy cow care, and will ensure a healthy udder through the entire lactation. No matter the amount, continue massaging and stripping until the teat is completely flaccid, and the udder yields no more than a drip of milk.

You can now open the cow gate to the milking stanchion, and allow the cow to slowly back out of the milking stall. She can return to her individual box stall, relieved of her full udder of milk, and lay back down to continue peacefully chewing her cud. Once all the cows have been milked, we let them all back out to pasture together, as a herd. We open the main gate to the barn, and then open their individual gates one by one, typically in the order that the cows were milked. They will orderly and eagerly follow one another back out to pasture.

Difficult Milk Cows

Sometimes a cow can be difficult during milking. This is usually because her udder or teats are very sore, and the milking process is painful. It is a natural response for any animal to try and avoid pain. Cows do this by moving away, or if constrained, by kicking. As such, a cow that is in pain or distress during milking will likely try to kick. She may kick at you, or at the bucket milker. Either way, we need to have a gentle yet effective method of correcting a kicking cow.

Many devices are marketed to stop cows from kicking during milking. This tells you that a kicking cow is not a particularly uncommon situation for a dairy farmer. There are cufflinks that hold the two hind feet together, and various contraptions that gently press into the cow's pelvis to discourage kicking. None of these products have ever worked satisfactorily for me.

The most effective solution for correcting a kicking cow is the old farmer's trick of tying back one hind foot. With one foot off the ground in a secure position, the cow cannot kick with that foot, nor lift the other foot to kick either. She is perfectly restrained, standing comfortably on one foot, in a position where she cannot kick.

It is important to attach the rope to the cow's ankle with a small belt. If you were to simply tie a knot, the force of the cow's kicking would tighten the knot so hard that you would never be able to untie it after milking. A wide belt is attached to the cow's ankle, and then a rope is tied to the belt. Gently pulling up on the rope, the cow will lift her hoof. You then secure the rope to a metal bar on the milking stanchion gate, in a position where it cannot move either forward or down. It is best to use a metal clip or carabineer to attach the rope, so that there is no knot on this end either. Simply clip the rope to itself, and when you are finished, unclip the rope, and release the belt around the ankle. Always be especially careful to not let your fingers get pinched by the rope when tying or untying the restraint.

Kicking is a learned behavior for the cow, so it is best to stop this bad habit as soon as it starts. Cows can kick with a tremendous amount of force, injuring the farmer or damaging milking equipment. As soon as the cow gets the idea that kicking might solve her problem, action needs to be taken immediately. Once gently restrained, the cow may try to kick a time or two, but should quickly recognize that nothing is gained by kicking.

Generally kicking problems will arise during the first milkings after freshening. This is particularly the case with a first time heifer. The engorged condition of the udder and teats can make milking painful at first for the cow. We have to have confidence that the best way we can ease the cow's discomfort is by milking her to relieve the pressure in her udder. She may not understand this, and may be distressed by the pain she is experiencing. As good animal stewards we sometimes need to do things that are uncomfortable for our animals in the short term, but in their best interests in the long term. Milking a painfully engorged udder is a perfect example of this principle of animal husbandry.

Teaching our Cows to be Good Milkers

Cows learn by watching one another. This is why we milk our top cow first. The other cows in the barn will have the opportunity to observe the milking routine of our best cow. They will be inclined to imitate the behavior they observe, especially as it is coming from the head cow in the herd.

This pattern of observation and imitation is especially important when we have a heifer that is just starting to milk for the first time. There are a lot of new things for the first time heifer to comprehend. Rather than trying to teach her everything ourselves, we give the young heifer the opportunity to watch and learn from her sisters. From the beginning of the milk season, the young pregnant heifer should be brought into the barn at milking, and given her own individual stall. She can slowly watch and assimilate the milking routine. She observes the order of events from the other cows, who she has learned to imitate in all other aspects of life. When the day comes to milk our young heifer for the first time, the routine will already be in her mind. We will have helped to make the learning transition as smooth as possible by giving our heifers plenty of time to learn through observation.

5. MILK HANDLING

Once we have milked our cows, there are many considerations for how we should utilize each day's milk. In addition to pure raw milk, the production of raw cream, raw butter, ghee, and raw milk yogurt provide a diversity of quality foods from the raw milk dairy.

Fresh milk direct from the cow is perfectly pure. Assuming our cow is of sound health and good udder condition, the milk we receive from our cows has an ideal microbial balance. Raw milk is a complex ecology of diverse beneficial bacteria, living enzymes, perishable amino acids, and fragile vitamins. Everything we do when handling our raw milk is carefully managed to maintain its perfection.

Filtering the Fresh Milk

Fresh milk will always pass through a disposable poly fiber milk filter immediately as it exits the belly milker can. This filtration will remove any large impurities that could taint the milk. Dirt or debris that gets into the milk briefly does not immediately spoil the milk. It is rather more like a tea making process, where the undesirable impurity, left to sit in the milk, will slowly dissolve and assimilate into the milk. Therefore, quickly filtering the milk prevents impurities from significantly impacting the milk quality.

The sealed design of the Surge bucket milker prevents 99% of debris from contaminating your fresh milk. However, it is still necessary to strain your milk through a fine dairy filter on its way into the bulk cooling tank. You should almost never see anything on the filter after use; it is more of a verification check for the cleanliness of your milk than an actual cleaning mechanism. Nevertheless, filtering your milk as you pour it from the milking machine into the bulk cooling tank is an important step.

The stainless steel milk filter functions like a large funnel, making it much easier to pour sixty pounds of milk into your bulk tank without spilling. The opening on the lid of the bulk tank is usually 6 inches in diameter, whereas the opening on the top of the milk filter will be more like 15 inches, making pouring much easier. Inside this large funnel, you will attach a disposable milk filter, which gets replaced daily. These poly fiber filters allow rapid filtration of the milk, while removing any micro contaminants. These replaceable filter disks are readily available from any dairy supply company, and cost only a few cents per day.

Raw milk, because it contains a diverse living ecology, is actually much less prone to bacterial contamination than conventional pasteurized milk. Raw milk contains its own defense antibodies and beneficial bacteria. When milk is pasteurized, all of these biological safeguards are destroyed, and the pasteurized milk becomes like a petri dish, vacant and awaiting bacterial colonization. Scientific experiments have demonstrated that when bad bacteria, such as e. coli are inserted into raw milk, they are completely destroyed by the living compounds in the raw milk within 24 hours. In contrast, the same e. coli bacteria, put into pasteurized milk, multiply rapidly and fully contaminate the pasteurized milk within a short amount of time. This is not to indicate that we can be careless with the handling of our raw milk, but to illustrate that raw milk is much more stable and safe than conventional agriculture would lead us to believe.

Cooling Milk to Storage Temperature

Cooling our milk to refrigerator temperature, between 34 and 36 degrees Fahrenheit, is the most important step for preserving the taste, nutrition, and microbial balance of fresh milk. We should have all milk cooled to this temperature within two hours of the completion of milking.

Healthy raw milk comes out of the cow in a state of optimal bacterial balance. Rapidly cooling the milk preserves this natural perfection by slowing metabolic processes of decay in the raw milk to a near standstill. Our milk is then stored in this stable condition that maintains flavor and nutrition.

A dedicated bulk refrigeration tank for cooling our milk guarantees excellent results. The combined effect of cooling the milk from all sides, with an agitator paddle to circulate the milk preventing any warm spots, ensures the best possible result. In the interest of maintaining the highest possible milk quality, a professional cooling tank is a necessary investment for a commercial raw milk dairy.

Once properly filtered and cooled, the primary source of contamination for our milk is ambient bacteria present in the air. For this reason, keeping our milk covered is essential. This is one of the many advantages to using a sealed belly milker system. Once our milk has been poured out of the belly milker through the milk filter, we want to limit air contact as much as possible. All milk holding containers must have lids.

One interesting and subtle detail is that as our milk is being cooled from cow temperature to refrigerator temperature, we do not want our lid to be absolutely airtight. During the cooling process, there are volatile compounds in the milk that will naturally escape into the air. We want a lid that can breathe slightly, so that these volatile compounds have an opportunity to escape. A perfectly airtight lid would keep the milk under slight pressure, which would hinder the milk's ability to freely release these volatile compounds into the air. We want our cooling milk to be able to 'exhale', without exposure to outside air.

Historically, milk was cooled in two to three gallon milk cans, placed in a large water trough. Ice from the icehouse, and cold flowing spring water were used to provide the necessary cooling mechanism. Another modern alternative to a bulk cooling tank would be a chest refrigerator, filled part way with water, with submerged milk cans. The biggest drawback to these alternative methods is that there is no stirring of the milk to prevent warm spots within the storage can. As such, when operating a commercial raw milk dairy, the use of a professional bulk cooling tank is highly recommended to guarantee optimal product quality.

In modern times, a refrigerator will not be adequate for cooling large amounts of milk at once, as would be the case if you did not have a bulk cooling tank. Refrigerators rely on a large thermal mass of cold items, and only small additions of warm food at any given time. Placing ten gallons of warm milk in a largely empty refrigerator will place a large strain on the refrigerator's cooling compressor. Additionally, the milk will take an unacceptably long time to come down to refrigerator temperature. In a pinch, especially for smaller quantities of milk, the best way to cool milk in a refrigerator is to utilize a large ice bath. For all things with the dairy, using the proper equipment ensures a job done right.

Washing Milk Handling Equipment

The unique combination of casein, lactose and butterfat make proper washing of milk handling equipment essential. Casein is the technical term for milk protein, and lactose is the term for milk sugar. The challenge of washing dairy equipment is that the casein needs to be washed off with cold water, and the butterfat needs to be washed off with hot water. Significant for our purposes, hot water will bake the casein onto our milk handling equipment, forming what is know as 'milkstone'. Cold water will harden the butterfat and prevent it from rinsing out completely. The solution to effectively washing dairy equipment is a proper sequence of cold and then hot water.

The proper system of washing milk-handling equipment is the same for all our equipment, regardless of whether it is made of glass, metal, or rubber. Our same sanitizing routine applies to the different pieces of the bucket milker, the bulk tank, our butter churn, glass jars, stainless steel milk cans, and utensils. Long handled bristle brushes, and specially shaped brushes for cleaning our inflations and other odd-sized pieces of equipment are standard cleaning equipment in the dairy. These dairy brushes are available from Hamby Dairy Supply in Missouri, and other dairy supply companies. Hooks hanging from the ceiling of the dairy, ideally in a spot where they receive direct sunlight, are an excellent way to store these brushes in a dry and sanitary manner.

When washing dairy equipment, rubber gloves are advisable. Gloves are more sanitary than bare hands, and the harsh chemicals in the dairy bleach will quickly damage the skin on our hands.

A lukewarm rinse to wash off the casein and lactose is the first step in sanitizing dairy equipment. We then follow with a hot water bleach scrubbing to dissolve any butterfat. The dairy bleach that we use is a specially formulated powdered detergent, which combines both soap and bleach in one. Dairy bleach powder can be ordered from Hamby Dairy Supply, and should be stored in a sealed glass jar to prevent caking from moisture absorption.

The final step in washing is a scalding rinse. This washes off any bleach residue, and leaves the equipment steaming hot, so that any moisture will rapidly evaporate dry. We now can organize the equipment on our drying rack, or if we are washing milk-handling jars, they are ready to be filled.

Sink, Counter, and Shelves

A large sink is necessary for washing up the milking machine, milk cans, the butter maker, and other dairy utensils. The sink should have deep and wide individual tubs, which can be filled with either hot or cold water. An integrated stainless steel counter that drains into the sink will be helpful for drying your dairy equipment. The best design for such a sink is a restaurant style, three-tub, commercial kitchen sink. These can be found at a good price when restaurants remodel their facilities, which is very common in the restaurant industry. The tub sink and counter is another piece of equipment where purchasing used is a very good idea.

With multiple sink tubs, an efficient washing process can utilize an initial cold bath, followed by a hot bleach bath, and a clean cold plunge. The clean utensils can then receive a final scalding spray, which will help the utensils to dry rapidly for optimal cleanliness. The freshly washed dairy equipment will then be laid out on the stainless steel counter to air dry.

The stainless steel counter portion of the sink makes an excellent work surface. This is where you can assemble your milking machine at the start of each milk session. This counter then works well for butter making, and organizing your containers for milk bottling. The integrated counter can be washed down easily, promoting excellent cleanliness in the dairy.

A wire drying rack will help you to organize your milking equipment during drying. After washing, place your drying rack on the stainless steel counter, in the sanitizing sunlight from your large window. Keeping your utensils carefully spaced allows air and sunlight to quickly and sanitarily dry your equipment. When not in use, your drying rack can be stored hanging from a pair of hooks installed in the dairy ceiling.

The best way to ensure organization in the dairy is with wire shelves, and hooks mounted to the walls and ceiling. Every individual piece of equipment needs to have its own proper home that it is returned to after drying. Hooks on the walls are good for hanging metal milk cans. Hooks hanging from the ceiling are good for storing brushes, ladles, and other small utensils. There are specially made metal hangers for your Surge milking bucket and inflations. A large wire rack can hold your milk filter and discs, latex gloves, butter churn, glass jars, and milk jugs.

Make a plan for all your dairy equipment, and where it will be conveniently and cleanly stored. There are a lot of individual pieces of valuable equipment that you will be responsible for in the dairy. Good organization and good cleanliness go hand in hand.

Bottling Milk

Once our milk has been completely cooled, it is time for bottling. For convenience sake, several days of milk can be combined in the bulk tank. With Monday's milk already cold in the tank, we can add Tuesday's milk, cool it to refrigerator temperature, and then bottle out both days milk at once. The automatic agitator paddle in the bulk cooling tank ensures that our milk does not separate into cream on top and skim milk on bottom. The milk in the storage tank is uniform in composition and temperature. Frequently, combining several days' worth of milk is more efficient than bottling milk one day at a time.

The first step in bottling is to ensure that all vessels to be filled are properly sanitized. When reusing glass jars or stainless steel milk cans that our customers have returned, we cannot assume that the vessels have been returned in a perfectly sanitary condition. Washing milk jars and cans is a time consuming process that we must do thoroughly every single time. If the vessels appear soiled in any way, a thorough prewash scrubbing with a brush and warm soapy water will be necessary. Hopefully, our vessels are returned in seemingly perfect condition. In this case, a standard washing of each vessel and lid, utilizing warm, dilute dairy bleach, is adequate.

With freshly sanitized vessels organized on our counter, we are ready to begin bottling. Utilizing the valve on the front of the bulk tank, combined with an elbow downspout, bottling into individual containers is easy. For large vessels like multi gallon milk cans, or wide mouth glass jars, the spout on the downspout will fit comfortably inside the mouth of the vessel. For smaller openings, such as are found on plastic milk jugs, a secondary stainless steel funnel will be needed.

Opening the valve slowly will allow the vessels to fill with a minimum of air bubbles, so slow and steady will actually be faster to completion than fast. If we fill too fast, air bubbles will end up in the milk in our vessel, and we will have to wait while these bubbles clear out of the milk to finish filling each individual bottle. Be careful to not overflow your vessel, as overflowing will create an unsanitary mess on the outside of the vessel, which will need to be thoroughly wiped down with a towel dipped in a dilute solution of dairy bleach.

A superior method for bottling large numbers of milk containers utilizes a large milk can, rather than filling individual milk bottles directly from the tank. Fill the large milk can, typically a three-gallon can, and allow any air bubbles to dissipate for a few moments. Individual vessels can then be quickly filled without pausing to allow bubbles to disperse. This is the preferred technique for filling large numbers of milk bottles at a time; the timesavings are significant.

Milk containers should be capped with an appropriate lid as soon as they are filled. Remember that air is our most common source of contamination. Sealing each vessel as it is filled ensures purity of product. There are specially produced plastic dairy lids for glass canning jars. These work well, and have the advantage of being a one piece design, in contrast to the traditional canning jar lid that had a lid and a ring, which both must be sanitized. Since the plastic of the lid is not in constant direct contact with the milk, concerns about plastic contamination are minimal. Ultimately, either style of lid for glass jars will work satisfactorily for our needs.

Once bottled and placed in the refrigerator, raw milk is remarkably stable. Of course, the perishability of your milk will depend on many variables, including the health of your cow's udder, your carefulness in handling, and the cleanliness of your farm and dairy environment. Assuming the best circumstances in each of these cases, raw milk in a sealed container in the refrigerator should keep for three weeks with no loss in quality. This is surprising to many, particularly those who have had raw milk from hand milked cows. Hand milked raw milk will always suffer more contamination, reducing its shelf life drastically. It is important to note however, that once the lid on the milk has been opened, airborne bacteria will immediately affect the quality of the raw milk.

We tell our customers that an unopened jar of milk will keep for two to three weeks, but that they should use the milk within a week after opening. In truth, we have experimented with placing a jar of raw milk in the back of the refrigerator, and have found the milk quality to be excellent after more than a month of storage. This is outside the boundaries of what we would recommend to our customers, but it does highlight the excellent keeping qualities of properly produced raw milk.

Choices in Milk Containers

Once your milk is cooled, it will need to be bottled for storage and delivery. As in all things, there are pros and cons to the different milk containers that are commonly used.

Stainless steel milk cans come in many sizes, with two and three gallon being the most useful. For handling large amounts of milk, these metal milk cans are ideal. The stainless cans are easily sanitized and reused; they handle transport well, and are durable. Their biggest downside is their awkward fit in refrigerators or coolers, and their initial expense. The best source for stainless steel cans is Hoegger Dairy Supply.

Half-gallon glass jars are the most practical size for milk transport. Gallon jars are too heavy, too breakable, and too expensive when they break. Smaller jars, like quarts and pints are excellent for yogurt and butter, but simply don't hold enough milk to be practical. Glass jars can be easily sanitized and reused, though careful handling is necessary to prevent breakage. When loading in a cooler or refrigerator, glass jars nest together nicely, making efficient use of the space. The best source for large quantities of glass jars is a local supermarket with an agreeable manager, who will be willing to special order you multiple cases of glass jars at a wholesale price.

The traditional plastic milk jug epitomizes convenience, and less wonderfully, disposability. Plastic jugs cannot be adequately sanitized for reuse, so customers will have to recycle their milk jugs, and you will be buying new ones by the hundreds. Sometimes, though, plastic jugs are the way to go, particularly if delivering long distances over bumpy roads, where glass would suffer far too much breakage. Sourcing plastic milk jugs affordably can be a challenge. The best arrangement usually comes from contacting local businesses that bottle their products in plastic, and seeing if you can get it on one of their large wholesale orders. You will likely need to order hundreds, or thousands of jugs at a time to be able to place an order. Due to the significant space required by empty plastic jugs, a large storage area will need to be planned out ahead of time. Unless stainless steel cans or glass jars are absolutely unworkable, plastic jugs are a poor choice for the farmer.

Refrigerator and Freezer

The milk cooling tank will do the job of cooling your milk down to a safe storage temperature. Once you bottle out your milk, or make butter or yogurt, you will need a cold place to sanitarily store your dairy products. Depending on the specific product, either refrigeration or freezing will be better ways of storage.

Refrigerator space will be needed for storing your bottled milk. Whether you bottle into glass jars or plastic milk jugs, a large amount of refrigerator shelf space will be needed to hold your milk while awaiting customers to receive their milk. Yogurt will need to be stored at refrigerator temperature as well. A large refrigerator with a good shelving layout will be critically important. Smaller refrigerators have a lower proportion of usable space inside, so larger units are generally much more space and cost efficient.

Freezer space will be less critical than refrigeration space. Often times a small freezer compartment within a larger refrigerator will be adequate. You will need freezer space to store any yogurt or cheesemaking cultures. The one dairy product for which freezing is the best storage method is butter.

Freezing Milk

For extended storage of raw milk, such as during the winter dry season, freezing provides the best solution. Frozen raw milk, when thawed, does not have the same texture as fresh milk. The unhomogenized raw milk will separate somewhat in the freezing and thawing process. This does not affect the flavor or nutrition of the milk, but the texture does become a bit grainy after thawing. For this reason, we do not consider frozen milk to be an adequate replacement for fresh milk, but rather an acceptable substitute during the brief winter dry period.

Milk cannot be frozen in large glass jars, as the expansion of the freezing liquid will crack the glass. If glass is deemed necessary, then wide-mouth quarts are the largest size that can be utilized. Freezing is one instance where plastic milk jugs really are much more practical storage vessels. Gallon jugs can be used, but are difficult to defrost; so half-gallon size plastic milk jugs are preferred for freezing large amounts of milk for storage.

When defrosting the frozen raw milk, allow the milk to thaw as slowly as possible, as this will disturb the texture least. The best way to slowly defrost frozen raw milk is to simply place the jug in the refrigerator, and let it thaw very slowly over the course of several days. To improve the texture of previously frozen raw milk, gently warm the thawed milk on a double boiler on the stove. The texture of previously frozen milk is dramatically improved by warming the milk to 110 degrees F, while gently stirring. Low heat, well below the pasteurization temperature, smoothes out the texture of the milk. For this reason, warm chocolate milk is a favorite wintertime treat for the dairy farm family.

Cream Handling

Production of raw cream is a simple process, as described in Chapter 3, in the discussion of the bulk cooling tank. Collecting several days' worth of milk is very efficient for gravity separation of cream. If we want the milk to separate for cream collection, we simple turn off the agitator paddle, and the refrigerated milk will naturally separate in the tank. Gravity separation of cream is an excellent technique for the small dairy.

Raw cream, due to its higher butterfat content, is more perishable than raw milk. Additionally, because the cream rises to the top, it is the layer of the milk that is in direct contact with the air, increasing its exposure to airborne bacteria. For these reasons, raw cream does not keep as long as fresh raw milk.

Raw cream, once bottled into a sealed glass jar, will typically keep for about ten days in the refrigerator. When the jar is opened and exposed once again to ambient air, the cream should be used within five days. This shorter shelf life encourages use of smaller vessels for holding raw cream. Pint or quart glass jars seem to be a good size, depending on the rate of use in the home kitchen. If you are providing raw cream to customers, it is important to educate them on the increased perishability of raw cream as compared to raw milk.

Buttermaking

The processing of large amounts of raw cream into butter is the most important method for preserving the essential nutrients in our perishable raw milk. Butterfat contains the lion's share of the nutrition in a given volume of milk. Butter from grass-fed cows is an absolute superfood for human health. The human brain is composed primarily of fat, so consumption of pasture butter feeds our minds as well as our bodies. Freezing butter for long-term storage is a very efficient use of freezer space, much more so than storing fluid milk. For these reasons, plus its self-evident deliciousness, making butter is an essential skill for the raw milk dairyman.

Churning your own raw butter is the best part of running your own small dairy herd. Dinner guests will long forget the deep conversations and lovely tablecloth, but they will drift off into a deep dreamy state of remembrance when thinking about the fresh butter you served. Farm fresh raw butter is the stuff of legends.

Fortunately, a perfect size butter churn is commercially available at a reasonable price. The Gem Dandy Butter Churn comes with a 3-gallon glass jar, motor, and churn. Fill the glass jar only about half full for butter making. As such, you will need 1.5 gallons of cream per batch, which will require roughly 8 gallons of fresh milk. This batch size should yield just over 2 pounds of fresh butter. If you separate off the cream when you come down to the dairy to begin milking, you can then churn several batches of butter while you are milking your cows. Each batch of butter will take about thirty minutes to churn. These individually churned batches of butter can then be combined, and all washed and salted together at the end, yielding quite the mountain of butter.

In time, you will inevitably drop the large, heavy, slippery glass butter churn jar. It will break. Do not worry; there is a great solution. When you are first learning to churn butter, the clear glass jar will be a real help for observing the butter making process. In time, you will be able to tell how your butter is progressing just by the sound of the motor. The visibility of the clear glass will not be important anymore. So about this time, when you break the glass jar, you can replace it with a three gallon, stainless steel milk can. Simply drill three holes in the can lid, and mount the butter churn motor and wand into the stainless steel lid. You now will have an unbreakable butter churn, with a nice strong carrying handle as a bonus. Breaking the glass jar will result in an upgrade, and you will never look back to the days of heavy glass butter jars.

The other upgrade that could be made to the Gem Dandy Butter Churn is to replace the plastic agitator wand with a stainless steel one. During the off-season, you could bring the plastic wand to a local metal fabrication shop, which could easily and inexpensively fabricate a stainless steel replacement. This would be permanently indestructible, and more sanitary than food grade plastic. Be sure that your metal fabricator understands the importance of smooth welds to ensure ease of sanitization. Now you have a first class, stainless steel butter churn, and the best butter your family and friends have ever tasted.

Cultured butter is a popular technique used in butter making. To make cultured butter, milk is first pasteurized, and then inoculated with a lacto culture, similar to that used in yogurt making. Since we are working with raw cream, which already has a full compliment of living cultures, this process of re-inoculation for making cultured butter is unnecessarily redundant. Customers or friends may ask if our butter is cultured, and the answer, is yes. Raw butter is naturally cultured with the indigenous microbiology that lives in raw milk.

Once we have collected our cream from the bulk cooling tank, we need to slightly warm the cream for optimal buttermaking. Remember to only fill the churn half way with cream. During the churning process, air becomes incorporated into the cream, increasing the volume of the original cream. Regardless of the batch size, we can warm all of our cream at once, and then churn individual batches. As our butter churns, one by one the tiny sphere globules of butterfat bounce into one another, and stick together. Warming the cream allows the butterfat globules to conglomerate more quickly. The ideal temperature for the cream is between 60 and 65 degrees, or roughly room temperature in the dairy. The best way to warm our refrigerated cream to this temperature is by submerging our stainless steel can of cream into a sink full of warm water. Body temperature water, about 100 degrees, will warm the cream at a good rate, neither too fast nor too slow. Any time that our butter does not want to consolidate during the churning process, temperature is the most likely cause of failure.

The churning of raw cream should take roughly 20 to 30 minutes. At this time, there should be a mass of yellow butter floating on top of the remaining buttermilk. Butter can be over-churned, where the mass of butter starts to get broken back up into smaller pieces, disappearing back into the buttermilk. For this reason, we need to monitor the butter, as it is churning, and stop the churn once we have a consolidated mass of butter floating on the surface. Excessively long churning times will result in less butter, not more.

Reach into the butter churn, and lift out the floating butter by hand. Place each handful of butter in a large, stainless steel mixing bowl. The friction of the butter churning will have further warmed the butter from when we started. As such, the butter we scoop out will be quite soft and warm. Enjoy this remarkable texture on your fingers, the sweet buttery aroma, and the brilliant yellow color. It is amazing to consider that this butter was derived, one molecule at a time, from the minuscule fats present in the grasses our cows were grazing. The enormous power of nutritional concentration, to convert plant leaves into butterfat, is a tremendous achievement by our precious dairy cows.

The next step is to wash the butter. The colder your butter washing water, the better the butter. Partially fill the mixing bowl with the butter, and then add ice-cold water. Squeeze your butter, noticing the whitish buttermilk that comes out of the soft butter mass. Drain off the water and buttermilk, being careful to not lose any butter, and repeat. Each washing only takes fifteen or twenty seconds. With each passing round, your butter will get darker in color and firmer in texture. Less and less buttermilk will drain out of the butter. Continue washing with fresh cold water; be very thorough in this step. When this step is completed, your butter will have cooled considerable. The mass of butter will be firm and waxy to the touch.

The final step is using the grooved, wooden butter paddles to extract any last remaining bits of buttermilk from the butter. Be sure to keep the mass of butter cold during this step, by periodically resting the butter mass in a bowl of ice water. Knead and fold the butter between your paddles, one small piece at a time. The process is similar to kneading dough for bread making. Any last remnants of buttermilk will be squeezed out of the butter, and will drain away through the grooves in the butter paddles. As each small piece of butter is completed, you can place it in a second bowl. This is finished sweet butter, and it is ready to eat.

Salting the butter is an optional, but traditional step in the buttermaking process. Gradually sprinkle salt onto the mass of butter, a little bit at a time, kneading the butter to work it in thoroughly. A good starting point for the salt quantity is one teaspoon of salt per pound of butter. The type of salt that you use can affect this ratio, so some experimentation is needed. Salt both helps to preserve the butter and enriches the flavor. Traditionally, sweet butter is used in deserts, and salted butter is used in all other culinary applications.

Ghee

Butter to be used for cooking, especially frying, is best clarified into ghee. Ghee is a traditional Indian method of preparing butter that cooks out any remaining buttermilk, leaving behind pure butterfat. In this condition of purity, ghee is a perfectly preserved, totally non-perishable product, requiring absolutely no refrigeration. With a substance this divine, it is no wonder that Hinduism holds the cow to be a holy deity.

To make ghee, place several pounds of butter in a large, thick bottomed, stainless steel pot. On the lowest possible heat, warm the butter until it is completely melted. Skim any foam off the surface of the melted butter and discard. Turn off the heat, and let the butter cool completely. Now place this pot of melted butter in the refrigerator for several hours.

The melted butter will solidify in the refrigerator, and any remaining buttermilk will separate to the bottom of the pot. Scoop off the hardened butter that is on top. Place this butter in a clean, thick-bottomed stainless steel pot and heat on your stove's lowest temperature setting. Milky colored bubbles will rise to the surface, which is good. Allow the melted butter to begin sputtering and popping, which happens as the last bits of buttermilk in the butter are evaporated. In time, no more bubbles will be seen. Be careful to not burn the ghee, keeping the heat extremely low, and removing the pot from the heat once no more bubbles appear. There may be a thin crust of dried casein on the surface of the ghee. This should be scooped off and discarded. Your ghee is finished, and can be poured through a wire strainer into glass pint jars for storage.

Depending on the ambient temperature in your kitchen, ghee at room temperature may either be a soft solid consistency, or a golden liquid. The melting point of ghee is around 75 degrees F, so its consistency and appearance can change significantly. This has no bearing on its keeping quality or perishability; it is merely an interesting phenomenon to observe.

Ghee is a remarkable cooking oil. It does not turn rancid when heated to high temperatures, unlike most modern cooking oils such as canola, olive, or soybean oil. Ghee has a very high smoking point, so it can be heated extremely hot for frying without any burnt flavor. From a nutritional and culinary standpoint, ghee is the absolutely supreme oil for all cooking applications. In baking recipes that call for butter, the different texture of ghee will not yield the same results. Similarly, when spread thick on a slice of bread, ghee will not have the same mouth feel as traditional salted butter. But for sautéing, frying, or braising, ghee is a culinary luxury that produces the best and most nutritious food.

Butter and ghee are best stored in pint glass jars, filled nearly to the rim, and sealed with a clean lid. Butter can be frozen for long term storage, and defrosted in the refrigerator when needed. Ghee is already completely preserved, and can be stored in a dark corner of the pantry. Of course, given the value of the ghee, it can also be kept in the freezer, where it will maintain its freshness and nutritional quality for years. A small jar can be kept safely on the kitchen counter, ready for immediate spooning onto rice or for frying an egg.

Raw Yogurt

Raw milk already contains a wonderful diversity of live cultures, which are beneficial to human health and digestion. We can increase the microbial benefits of raw milk by further adding the cultures of raw milk yogurt. Yogurt is traditionally made from pasteurized milk, where the native microbes in the milk are destroyed through heat treatment, and then a new set of microbes are added to the pasteurized milk. When working with clean, fresh, raw milk, this is unnecessary. We can enjoy the best of both worlds, the microbes present in raw milk, and the additional beneficial bacteria of yogurt cultures.

A household size pasteurization machine makes an excellent yogurt maker. Anytime you are heating milk, indirect heat is required. Indirect heating is accomplished by using two independent vessels to create a water bath. The water in the outer pot is heated, and the heat is indirectly transferred into the inner pot to warm the milk. Indirect heat allows the milk to be warmed slowly, without any scorching of the milk. Scorching ruins the delicate flavor of the milk, and compromises the structure of the fragile milk proteins. For yogurt making, a small capacity pasteurization machine will allow carefully controlled heating of the milk with indirect heat.

A small pasteurization machine controls the heating of the milk with a thermostat. This way, our milk can be brought perfectly to a set temperature with no monitoring. Once the milk has warmed to the pasteurization point, a switch will allow you to engage a lower secondary temperature set point, so that the milk can be held at the precise temperature for incubating your yogurt.

The small pasteurization machine sold by Hoegger Supply will process 3 gallons of milk at a time. You can ensure perfect quality control in your yogurt making process, with the thermostatic controls. For home use alone, smaller batches of yogurt can be made on your stovetop, with common household pots. But for small commercial applications, the precision of a small batch pasteurizer is a recommended upgrade at a reasonable cost.

Gourmet yogurt cultures can be purchased in a freeze-dried, powdered form. Glengarry Cheesemaking Supply is a good source for top quality yogurt cultures, typically imported from France. We use this pure powdered culture to make a homemade, traditional yogurt mother. The advantage to using a yogurt mother for inoculating our raw milk yogurt, is that it is much more economical. Freeze dried yogurt cultures, imported from France, are expensive when you are making large quantities of yogurt. Making a yogurt mother allows us to multiply the utility of a given quantity of yogurt cultures by a factor of several dozen.

The best results are achieved when we make our yogurt mother in the traditional way, involving pasteurization. We heat our milk to 145 degrees F, for 30 minutes, ideally utilizing the small batch pasteurizer described above. Alternatively, if such an appliance is not available, very carefully heat the milk on the stovetop, using the indirect heat of a double boiler. Once the milk has been properly pasteurized, allow the covered milk to cool to 110 degrees, and then add the dried yogurt cultures, carefully following the directions on the package.

When making a yogurt mother, it is imperative to use utensils that a perfectly clean and properly sterilized using either boiling water or dilute dairy bleach. The yogurt culture is stirred into the hot milk in either pint or quart glass jars, this step is called inoculation. A secure lid is sealed on top immediately to prevent any airborne contamination. The sealed jars of freshly inoculated yogurt mother now need to incubate at 110 degrees, for four hours.

Incubating yogurt, whether it is a pasteurized yogurt mother, or raw milk yogurt, is most easily achieved in a cooler filled with hot water. Place the glass jars of incubating yogurt in a cooler, and fill with 110-degree water. Close the lid, and let the yogurt mother incubate for four hours. The cooler will hold enough heat for the incubation to work flawlessly. After four hours, remove the jars of yogurt and place in the fridge. It is important for the yogurt to cool quickly and be stored at refrigerator temperature.

The finished yogurt mother is now used to make raw milk yogurt. We use warm milk, fresh from the cow, and add a tablespoon of yogurt mother to each quart jar of raw milk. Pint jars require less yogurt mother. Make sure that your jars and lids are sterilized properly. Stir in the yogurt mother with a sterilized utensil. The process of making raw milk yogurt requires a longer incubation than for yogurt mother. Keep your jars incubating in a cooler full of 110-degree water overnight, for about 12 hours. The water will cool slightly during this longer incubation period, but that will be fine.

Sealed jars of yogurt mother will keep in the refrigerator for several months. It follows that one large batch of yogurt mother, filling a dozen individual jars, will last for months worth of raw yogurt making. Once opened, the yogurt mother cannot be resealed for future use. Raw milk yogurt in an unopened jar will keep for a full month in the refrigerator. Once opened, the raw milk yogurt should be consumed within a week.

At the end of the milking season, in anticipation of the winter dry period, we typically make several large batches of pasteurized milk yogurt for long-term storage. We pasteurize the milk for the yogurt, then inoculate with the yogurt mother, and incubate for 4 hours at 110 degrees. Pasteurized milk yogurt in a sealed jar will keep for six months, actually improving in flavor and texture during storage as the microbial cultures fully digest and transform the nutritional compounds in our milk.

During the milking season, we make all of our yogurt as raw milk yogurt, because it is less energy intensive to produce, is a unique artisanal dairy product that our customers cannot get elsewhere, and has superior nutritional qualities.

Cheesemaking

Small batches of cheese can be made for the farm family using the small pasteurizer/yogurt machine. In general, you can figure that each gallon of milk will make just under a pound of cheese. Cheesemaking is a complicated and skilled artisanal craft, which deserves a book of its own. Given that cheesemaking is generally not an appropriate component of the small dairy, this book will not go into detail on the equipment or techniques of cheesemaking. Those interested in cheesemaking are advised to consult Ricki Carrol's book, *Home Cheesemaking*, and Margaret Morris' book, *The Cheesemaker's Manual*.

6. ESTABLISHING THE DAIRY HERD

Pasture based dairy farming requires the right animals for success. Our expectations for our cows' performance, and our management methods to achieve those goals are very different from conventional dairy operations. We need to utilize dairy cows that will excel in this more natural, holistic system of dairy management.

The Ideal Dairy Cow

The perfect cow is strong, athletic, and beautiful, with a healthy udder and delightful disposition. Extreme milk production records are not our objective. Our cows need to be able to maintain healthy body condition, evidenced by a slick and shiny coat, all through the year. They will reliably conceive during their first heat with the bull. Their pregnancies are uneventful and their births occur naturally without assistance or complication. Our cows' milk production should be consistent, not requiring any feed supplements or veterinary intervention. This bovine is naturally adapted to the life of a pasture dairy cow. She and her offspring are priceless to the small raw milk dairy.

The Breed Spectrum

There is a broad range of genetic types within the bovine species. Cows are not truly natural animals. They have been bred and selected to a degree that, like corn, they are an essentially human creation. There are no wild herds of cattle that we can look to as the 'original' animal type. The last wild cattle went extinct well over a hundred years ago in Poland. The dairy cows we raise today are products of human management and direction. We can take great pride in stewarding such a treasure of human agricultural heritage.

Cows breeds, and individuals within those breeds, exist along a spectrum that ranges between two extremes. At one end, we have beef breeds, like the continental Simmental and the British Angus. These animals gain weight with every bite of feed they ingest. They have been selected either for maximum rate of gain, in the case of the Angus, or maximum size, in the case of the Simmental. Their sexuality and milk production is diminished, as all focus has been on developing their body mass for meat.

At the other end of the spectrum are the specialized dairy breeds, like the continental Holstein and the British Jersey. These animals grow body frames with a minimum of musculature and body fat. They have been selected for either maximum milk yield, in the case of the Holstein, or maximum butterfat content, in the case of the Jersey. Their breeding has selected for individuals that place all their nutritional resources into milk production.

Both of these cattle types are 'extreme', in that they are far to the ends of the breed spectrum that farmers have developed over the millennia. Certain traits have been accentuated at the expense of the overall balance of the animal. These breeds are highly adapted for their specialized tasks, but selection for such specific traits has come at a price. Beef breeds have very low milk production, and dairy breeds have overly thin body condition. A holistically managed pasture dairy needs animals that are more balanced, and combine some of the more desirable traits from each end of the breed spectrum. What is needed is a 'dual purpose' animal.

In between the extremes of pure dairy type and pure beef type, we have what are referred to as 'dual purpose' cattle breeds. These breeds include Ayrshire, Brown Swiss, Milking Shorthorn, and Dexter. In essence, these breeds combine a thicker body type with good milk production.

The genetic character of dual-purpose breeds does not direct their entire nutritional intake into milk production. They still are good milk producers, but will also store away thick body reserves for leaner times. These nutritional reserves are essential for maintaining optimal health when raised exclusively on pasture. The Jerseys and Holsteins, in their modern form, require a perfectly tailored diet every day, as they carry no bodily reserves. They have been managed for generations to require human balancing of their nutritional intake; otherwise their frail bodies rapidly become depleted and their health strained. The robustness of the body of dual-purpose cows acts as an insurance policy against nutritional deficit.

Milk quality, as measured by butterfat and milk solids, has significant variation among breeds. Holsteins are notorious for their high fluid volume, but low fat and solids content makes their milk inherently less nutritious and tasty. Jerseys are equally renowned for the high butterfat content of their milk. Unfortunately, butterfat is the most nutritionally demanding part of the milk for the cow to produce, and combined with their lack of bodily reserves, is a major factor in why Jersey cows frequently are unable to maintain healthy body condition when raised exclusively on pasture. Their small frame size and smaller digestive capacity make it problematic to balance their butterfat output with their nutritional intake. Balance is key in holistic management. The dual-purpose breeds are more balanced in their milk composition, and more robust in their body types. These two factors generally promote healthier milk production in a holistic, pasture-based dairy.

Among the dual-purpose breeds, milk quality is generally excellent. These breeds all produce a good amount of butterfat, generally between four and five percent, and also high levels of milk solids. This makes the milk from these breeds delicious for raw milk, excellent for cheese production, and great for butter making. More detailed statistics on butterfat and solids levels by breed are really just statistical averages, as individual animals will demonstrate significant variation. There is no need to get caught up with the idea that one breed might have a percent more butterfat than another, given this individual variation. The milk quality from dual-purpose breeds is excellent in every way.

Dual-purpose cows received that designation because they are capable of being raised profitably for both milk and meat. In the pasture-based dairy, we utilize their more robust physical condition as a buffer against nutritional deficiency. The dual-purpose cow can metabolize her body reserves to maintain optimal health through periods of slight nutritional stress. It is a bonus that our surplus male animals also develop enough muscle to produce substantial beef animals. The fifty percent of our calves that are male will have good beef yields with excellent meat quality. In the holistically managed herd, the economic value of this beef will be significant on a yearly basis.

Deciding on a Breed

Our search for the ideal cows for our small dairy begins by narrowing down the breed options. We should likely discard extreme types, such as Jersey and Holstein, and focus on the more versatile dual-purpose breeds. These breeds themselves exist along a spectrum of traits. Brown Swiss are very large cows, with mature females reaching 1400 pounds. Milking Shorthorns and Ayrshires are more moderate in size, with cows maturing to 1100 pounds. Dexters are considered a dual purpose breed due to their balance of milk and muscle production, but are very small animals, reaching just 800 pounds. Total weight aside, these breeds should all maintain good body condition while being milked on pasture alone, and are ideal for a holistically managed dairy.

Milk production generally will be proportional to total animal size. On a very limited acreage, a farmer could keep almost two Dexters for every one Brown Swiss. The milk yield would be similar between the two scenarios. The biggest consideration is that the milking labor would be almost double, dealing with two Dexters as opposed to one Brown Swiss cow. If herd size based on land area is the biggest limit to one's dairy potential, this tradeoff may be completely sensible, despite the increase in labor requirements for the farm.

Brown Swiss are massive cows. They have a very rugged physical constitution, and handle both heat and cold with ease. Their milk is notably high in milk solids, as is regarded as the best milk for yogurt and cheesemaking. They combine a solid body condition with excellent milk production traits. Milk and butterfat yields from Brown Swiss cows can compete with the very best dairy cows in the world. They have solid feet and strong bodies, tending to be very vigorous and healthy animals. Their beef potential is excellent, and their large stature benefits raising them for oxen. Brown Swiss cows are reasonably well distributed across the country, and are readily located for purchase.

Ayrshire cows originate in Scotland. They are moderate size cows, with beautiful upright horns. They are much better adapted to cold and wet environments than hot and dry. Ayrshires are regarded as particularly thrifty cows when raised on pasture. They are moderate size cows, with excellent dairy production qualities. Their biggest drawback is their rareness, which may make sourcing Ayrshire cows difficult in many regions.

Milking Shorthorns are a truly American breed of dairy cow. They are well rounded in all characteristics. They can be raised effectively as dairy animals, for bull beef, and as working oxen. Their dairy production tends to be somewhat less than Brown Swiss or Ayrshire. Despite originating in America, their numbers have always been small. They have always been regarded as an excellent homestead cattle breed, so if they are available in your region, they merit serious consideration.

Dexters are considered miniature cattle, with historical origins in Ireland. They are very efficient grazers, and combined with their smaller size, make excellent use of smaller pastures. Their temperament is very gentle. Their milk production can be quite good, although significant variation exists within Dexter herds, and not all individuals express optimal dairy qualities. The biggest consideration with Dexters is a question of if you want such a small animal as the basis of your dairy herd.

Local availability will be the biggest single factor in determining which breed is best for your farm. If there is a local Brown Swiss dairy, with healthy and holistically managed animals, that would likely be the best way to establish your herd. A local or regional dairy will have animals better adapted to your climate and soil conditions. You will be able to examine potential animals more thoroughly. Costs for hauling will be minimized. Start your search nearby, and only consider far-flung options if you cannot find good animals close to home.

Purebred Cows

From a practical standpoint, we do not need purebred cows to develop a healthy and productive cowherd. Purebred cows will be easier to evaluate, because we know from examining the breed standards the proper conformation of a given animal. However, there is virtually no value in 'registered' purebred animals. Production quality animals are our criteria, and productivity has almost nothing to do with an animal's registered pedigree.

We are running a productive farm, not a show cow operation. Registered animals will cost more, due to the cost of maintaining their registration. These animals are generally not worth their price premium. For our purposes, we will want to keep our best offspring to improve our cowherd, so the premium that we might be able to charge when selling registered animals is not a significant part of our farm economics. The world of purebred registered cows is similar to that of dog breeding or horse showing. They operate by their own rules, which have no bearing on animal productivity on a small farm.

Crossbred dairy cows of mixed breed heritage should not be discriminated against in the founding of our herd. First and foremost, we want cows that meet the specific performance objectives described at the beginning of this chapter. The color of their coat is of no importance to their value to us as productive farmers. Hybrid vigor is a very real concept, and as such, mixed breed animals may even have certain advantages. Of course, the specific pedigree of a given individual must make sense. For example, crossing a Brown Swiss to a Dexter would make no sense, and would result in inferior offspring. But crosses of Ayrshires and Guernsey, or Brown Swiss and Milking Shorthorn, could certainly result in excellent quality dairy animals.

All things equal, finding purebred, but unregistered, cows will likely be our most desirable situation. Working within a given breed will make our evaluations of individuals easier, and our selection of breeding bulls more straightforward. We will have established criteria to aim for in the development of our herd's conformation. Selecting bulls to improve our herd's shortcomings will be much easier when we have a narrower genetic base from which to establish our breeding program. In all regards, breed consideration is secondary to the quality of the individuals in developing a productive small dairy herd.

Finding the Right Cow

Visit as many dairy farms as you can, observing and evaluating different cowherds in their home environment. There will be some farms that manage their animals so differently to your holistic objectives, that you would be wise to not give them much serious consideration as a buyer. The specific breed is not as important as the way that the animals are managed.

The most problematic situations occur with animals that are accustomed to being fed large amounts of grain. These cows' rumens will be maladapted to a pasture only diet, and will struggle to maintain health and productivity in a radically different feeding system. Many conventional dairymen do not feed their heifers any grain until after they have calved, so in this situation, purchasing heifers when young can mitigate against the destructive effects of feeding grain to cows.

There may be an organic, holistically managed dairy, raising top quality dual-purpose animals in your area. That would be ideal, surely. Expect to pay a real premium for these animals. The vast majority of dairy cows in this country are not being managed holistically, and those that are,

are both rare and valuable. Be sure that the cows you are examining are sound in health and temperament. There is a well-founded saying in the dairy world that no good farmer ever sells a good dairy cow. A good dairy cow is worth far more in her productive value and the value of her future calves, than any price she might fetch. As buyers of dairy cows, we are almost invariably stuck with buying somebody's rejects. The important thing is to understand why a given cow is being sold from somebody else's herd.

Accepting that we will likely not be buying a perfect cow, we must consider what factors are most critical in our buying criteria. Reproductive performance comes first. The cow must have a perfect history of conception and calving. Shortcomings in fertility will turn a potential dairy cow into a feeding liability. She may be inconsistent in conceiving, costing us valuable lost production during pasture season, and compromising our seasonal calving schedule. She will still eat tremendously, but not repay us with her milk production. Cows must have a perfect record of conception and healthy calving to be considered for a holistic dairy.

The health history of your potential cow comes next. Particularly concerning are any health issues that arose as a calf, especially as a result of inadequate colostrum intake or milk supply when young. A stunted calf will never develop the maximum size digestive system that is needed for robust milk production. Parasite issues as a calf will permanently compromise that animal as it grows into maturity. The inner workings of a cow are like a complex ecosystem, and like repairing damaged forest ecology; it is a very long and slow process to heal a degraded microbial ecology in a mature cow. Calves must develop perfectly to become good producing milk cows themselves. Ask lots of questions about a cow's health history, especially as a calf and during her first pregnancy.

Milk production records should be consulted for any dairy cow you are considering purchasing. Any dairy farmer worth buying from will have detailed records of the daily milk production of all their cows. Any lapses in production from mastitis are immediate concerns. Large variations in milking yields through the milk season are not good. We want cows with excellent persistence, rather than extreme peaks. Persistence is the ability for a cow to produce a high and sustained yield of milk through her lactation. Of course, all cows will decline as their lactation progresses, but the more consistent her production is, particularly after conceiving, the stronger her constitution, and the more profitable her management.

Comparing individual cows within a given dairy will give you a good head to head comparison of individual performance. Specific production numbers for milk yield must be taken in a relative context, as your management will be different, and your yields will accordingly not be the same.

Genetic Weakness vs. Behavioral Weakness

Establishing a new dairy herd will generally require utilization of less than perfect cows. As stated previously, no dairy farmer will sell you their best cows, at least not at a price that would be economically viable for a startup dairy. Accepting that our initial animals will have some flaws, an important distinction is whether the flaws are genetic or behavioral.

Genetic flaws will be passed on to all future offspring. Inherited traits, such as udder health, body conformation, and reproductive performance, are factors that will be problematic in this individual cow, and in all of her offspring. Animals exhibiting such critical shortcomings should be avoided, if at all possible. These are terminal problems that will never go away, and will be a curse upon all future generations.

Other genetic traits, such as coat color, tail length, or facial appearance, are not particularly critical for the needs of the dairy farmer. This is to illustrate that not all genetic traits are equally critical, although they all are rooted in the biological reality of the animal. Buying an ugly cow, who has otherwise sounds production qualities, would be a wise purchase.

Cows with behavioral shortcomings are the type of animals that should be looked at closely for a savvy bargain. Many times a given cow may have had a bad experience when young, leaving her with a bad attitude. Often times, a change of scenery will result in a complete change in disposition. The selling farmer may be relieved to be rid of a troublesome cow, who with a little careful rehabilitation, becomes a delightful milker on a new farm.

In a large dairy, a kicking cow may be very disruptive to the efficiency of a large milking line. Brought to a small farm, where a careful attention can be applied to break the bad habit, the undesirable behavior can be corrected, and a great cow can result.

Of course, distinguishing between a hopelessly troublesome cow, and a cow in need of a change of scenery, is a delicate art. Some cows are simply ill tempered, and are best avoided under any circumstances. Fortunately, this is an uncommon reality, as dairy cows have been carefully selected for so many generations, that good behavioral temperament has been deeply engrained.

It is important to distinguish between behavioral flaws and genetic flaws. In general, genetic flaws are a red flag for the new dairy farmer. Behavioral flaws can be remedied, in many cases, with careful attention and good animal husbandry. When searching for potential dairy animals, knowing to distinguish between these two categories of problem cows is an important consideration.

Three Types of Cow to Buy

One class of potential cows that merits serious consideration, are cows that cannot handle the social requirements of a large herd. These are the 'bottom cows', cows that get bullied and generally have a miserable time competing in a large herd for feed and space. Brought into a new environment, on a small farm with less competition and more space, they can absolutely thrive. Free from the hardships of constant bullying in a large herd, their production can rise, and suddenly they can become excellent dairy cows. You are taking a chance, of course. But these cows that struggle with the competition of a large herd can constitute a willing sale for the large dairy farmer, and a savvy purchase for the startup dairy looking to start with a quality genetic base.

Recognizing that good dairy farmers rarely sell good dairy cows, we may be inclined to consider purchasing heifers to start our herd. Generally, bred heifers are much easier to find for sale, and are more economical to purchase. Heifers are more likely to not have been fed any grain that would disrupt their rumen microflora. Heifers are inherently untested, and despite coming from an excellent dam, they may not have the genetic dairy quality that we desire. Heifer cows will produce less in their first lactation than in subsequent seasons, so their productive value for the farm will start off slow. We will not know their true productive value for years, until after several lactations have been documented, and significant time and resources have been invested.

Heifers will require training in the milking routine. When we have a balance of mature cows and heifers, the heifers have the opportunity to learn the routine by observing the other cows. On a dairy farm that is initially established with only heifers, the training process may be more challenging without any natural leaders. Our skill in handling and training large animals will be tested significantly when starting off exclusively with heifers in the milk stanchion.

Accepting that we likely will not be able to source or afford perfect cows in the peak of their productive life, finding older, soon to be retired cows, may be more realistic and practical. As cows get older, their production slowly starts to decline. The wear and tear of years of lactating shows up in increasing udder problems. Mastitis is not considered an acceptable problem, and those cows must be explicitly avoided. Frequently, the continual pull of gravity will cause the cow's udder to hang undesirably low, making milking more difficult. Milking will come to take more time in the stanchion that is acceptable in a larger commercial dairy. These cows are the kinds of cows that can often be purchased at a reasonable price. We can expect that with more caring management, an older cow will still produce well for us for several years, yielding valuable calves in the process.

A combination of an older, proven cow or two, and an untested heifer or two, may make an excellent foundation for a small dairy herd. The older cows will be docile and reliable, helping to train our heifers. The untested heifers will be slow to yield, but will be coming into their prime as we retire our older cows. We will be producing a good supply of replacement calves each year, establishing the future generation of our small dairy. The cost of purchasing heifers and older cows should be reasonable, enabling our dairy to produce a profit from the beginning. This is a holistically balanced situation that establishes our small dairy in a productive and regenerative way.

Long-Lived Productive and Profitable Cows

Our perfect dairy animal is a robust and healthy cow that produces a moderate amount of high nutrition milk. Health comes first; it is the underlying principle of our entire herd management. Conventional dairy cows, with their unnatural management, are constantly on the breaking point of physical wellbeing and illness. All of the philosophies in this book, with regards to seasonality, pasture feeding, and natural cattle care, are designed to keep our dairy cows in optimum health. Cows that enjoy and long and productive life are more profitable for the farmer. We invest a lot of time and resources into raising a calf to become a dairy cow, three years of commitment on average. Regularly replacing cows that have declined in health or production places an unsustainable financial strain on the dairy farm. Strong and healthy dairy cows can and should be good milk producers for a decade or longer. I have seen Brown Swiss cows that have had fourteen calves in fourteen years, producing milk for the dairy every step of the way. This happens when we combine good management with healthy cow genetics, building our dairy farm in a sound and sustainable way.

7. QUALITIES OF THE PERFECT COW

Objectively comparing our cows and selecting the best individuals is essential for the genetic improvement of our herd. Quantifiable factors, like milk yield and reproductive consistency, are easy to compare and evaluate. More subjective traits are less easy to judge, but nearly as important from a herd improvement standpoint. What follows is a discussion of the more subtle points that should be considered when evaluating any cow.

Udder Characteristics

The cow's udder should be covered in fine, downy hairs. Long hair growing on the udder is not a good sign. In winter, a cow may have longer hair hanging down over the sides of her udder to protect it from cold, but long hairs should not be growing out of her udder at any time.

It may be counter-intuitive, but udder size is not directly correlated to milk production. All things equal, a compact but high-producing udder is much better for both cow and farmer. When we first see a cow with a humungous udder, it is natural to assume that she must be a massive producer. This is not necessarily true. Udder size can have as much to do with distention and edema, as milk yield. If anything, be skeptical of the cow with a toilet-sized udder. Dairy farming will be more labor efficient, and less problematic, when working with cows that have compact udders.

A compact udder indicates less edema, or swelling, occurring within the lactation tissues. This will promote healthier milk, with less potential for mastitis. An overly swollen udder does not receive optimal blood circulation, because the blood vessels are constricted by the pressure caused by swelling. Without adequate circulation, the immune system cannot cleanse the lactation tissues as effectively to prevent bacterial infection. The end result is that overly engorged udders produce lower quality milk and are more prone to mastitis.

The negative effect of gravity on an oversized udder cannot be underestimated. A full udder weighs nearly one hundred pounds, and it jostles back and forth walking through the pasture, it is constantly straining its attachment ligaments. Compact udders exert much less strain on these ligaments, so the sagging effects of gravity are less pronounced. The cow with a compact udder will maintain much better udder condition and udder health as she ages more gracefully.

In the milk stanchion, a compact udder will more easily accommodate our milking can. The teats will be higher off the ground, so that calves will have an easier time nursing. As long as milk yields are comparable, compact udders have nothing but advantages.

The teats should be equal in size, and balanced in their distribution on the udder. Super-sized sausage teats are more prone to infection, and more likely to be damaged by being accidentally stepped on by a cow standing up in the pasture. Also, excessively large teats will not fit properly in the inflations of the milking machine. Teats that are too small will be problematic for hand milking, and more difficult for newborn calves to nurse.

Some cows will have an extra 'false teat' or two at the back of their bag. This is not a problem, and should not be any cause for concern. False teats that are located near the cow's real teats are undesirable, though not a fatal flaw. A good even distribution of the teats on the udder, located symmetrically, indicates a well-formed milk organ.

The ideal teat size is about the size of your ring finger, and slightly wider. The teats will shrink considerably when the cow is dry, so evaluating teat size at this time, or in a heifer, is more difficult. In general, the teats should all be the same size to facilitate easy operation of the milking machine.

Good Feet

The feet of our cows endure tremendous pressure with their small footprint and the massive weight of the mature cow. Broader feet distribute this weight more evenly, and are less prone to problems. Some breeds, like Brown Swiss, are particularly noted for their solid feet, while conventional dairy cows frequently suffer hoof problems.

The rate of hoof growth, and the rate of wearing away should be evenly matched. A pasture environment, where our cows are walking a lot, sometimes over rocks and stone, should maintain healthy feet with no intervention from the farmer. In environments where our cows are not walking much, or the ground is uniformly soft, hoof trimming will be necessary. Overgrown hooves will be prone to cracking and splitting, which increases the risk of infection. Occasional examination of the cows' hooves is important to address any minor issues before they become true problems.

The canon bone, located just above the cow's hoof, is indicative of the level of refinement in the dairy cow. Our cows should have slender, hourglass-shaped, canon bones. Large, club-shaped canon bones indicate an unrefined animal that is placing too much nutritional resource into excessive bone growth. Our cows should have large, broad feet with slender, feminine canon bones.

Wide Mouth

A grazing cow with a wide mouth consumes more feed with each bite of pasture. This cow will be able to spend less time working to nourish herself, and more time contentedly chewing her cud and producing more milk. Broad muzzles are valuable for pasture-fed cows.

The tiny droplets of moisture that appear on the cow's nose are actually microbial pasture inoculants. The microbial composition of these beads of moisture is similar to that of excellent compost tea. As our cows graze in the meadow, they are constantly stimulating the pasture plants by inoculating them with these beneficial microbial compounds. It follows that a cow with a broader muzzle has a larger area of these secretions, and accordingly spreads more of this beneficial inoculant on our pastures. Cows with broad muzzles graze more efficiently, and help our pastures to grow more healthily.

Sprung Ribs

The girth of our animal's ribcage is the limiting factor in determining the size of their internal organs. Cows and bulls that have large, barrel chests have a larger lung capacity, better circulation, and most critically, larger digestive systems. The legs of the animals should be widely spaced, to accommodate a larger chest cavity. The shoulders should be broad, with the shoulder blades level with the spine. This arrangement of the skeletal anatomy will allow for maximum chest girth.

Looking at a cow or bull from behind, their ribcage should spring outward, with a pronounced curve to the ribs. 'Slab sided' animals are undesirable, as their digestive capacity is reduced. Animals with well-sprung ribs can ingest more food, which is critical for animals that are deriving their entire nutrition from pasture grazing. Digestive capacity is strongly related to milk production, so we want to see animals with a large chest girth, as a means to maximum milk production and animal health.

Bright Eyes

The eyes on our animals should be clear and bright. Cloudiness in the eyes compromises their ability to see properly, and may be indicative of an underlying nutritional deficiency. Sometimes, cows and bulls can have their eyes damaged by getting scratched on coarse dried forage. To prevent this, we do not want to be grazing our animals in overly mature pastures, which inherently have minimal nutritional value, and pose a risk to the health of our animal's eyes. Dried seedheads are the worst offender for seriously scratching a grazing animal's eyes. Minute scratches on the cornea can easily become infected, leading to serious eye infections that can cost the animal their eyesight for life.

An animal with eye problems will have dried crusty debris around the eye and extensive tearing. When these signs are evident, prompt treatment is necessary. Bring the animal into the milking stanchion, where they can be gently restrained. Squeeze the pulpy juice from a fresh cucumber, and gently apply it directly to the eye. Repeat this treatment daily, or even twice daily, until all crusty discharges are eliminated. The cucumber juice treatment astringes the surface of the eyeball and squeezes out any foreign material. The cucumber juice is also soothing and moisturizing, and is a remarkable remedy for any eye problems in your herd.

The Female Form

The ideal female form is graceful and refined. The neck and tail both should be long and slender. These features indicate refinement in the female bovine form. Our females should have small heads, with graceful and well-proportioned facial features. The dairy cow is the embodiment of femininity, a living deity, and should have a beautiful appearance.

The pelvis and rump on the cow should be as broad as possible, to facilitate ease of calving. The width between the two pin bones should be great. A wide stance in the hind legs will both make calving easier, and will make supporting the weight of a full udder more natural. The ideal dairy cow should taper significantly from her rump to her shoulders, with a distinct wedge-shaped appearance when viewed from above.

The horns on the female cow should have a smooth and almost lustrous surface. There should be no noticeable rings of discoloration on the horn, which would indicate times of nutritional deficiency. On a mature cow, you may notice slightly raised ridges on the horn, which correlate with the stress of calving. Like reading tree rings, in some cows we can read the annual history of their reproduction in the raised rings on their horns.

The form of the horns varies by cattle breed. There are no general guidelines to proper shape that apply to all cows. In any case, the horns should be symmetrical in form, and balanced in proportion. Occasionally, a cow may damage a horn when young, rendering it crooked, and this should not be considered a flaw. A crooked horn will continue to grow normally, and will often straighten itself out over time.

The cow's horns are functional organs of perception. They are filled with blood vessels and nerve endings, and clearly perform some important function for the cow, that is yet unknown to science. It may be that the cow horns are used for communication, with one another, or with the universe at large. The mystery of cow horns is wonderful and intriguing. Well-formed and beautiful horn structures are indicative of healthy and vibrant cows.

Breeding Bulls

The dairy breeds have highly accentuated sexual characteristics, which promote maximum milk production in the cows. An interesting and important phenomenon is that more masculine bulls tend to produce more feminine female offspring, which is our ideal in developing our dairy herd.

The ideal male form is powerful and massive. The breeding dairy bull is a hyper-masculine animal, thick in stature and imposing in form. In distinct contrast to the females, our bulls should have short and stocky necks and tails. The head should be large and blocky. Maximum girth in these areas indicates high testosterone levels and optimal fertility.

The general physical form of the bull is similarly reversed from the ideal cow. The rump on a bull should be slender, surprisingly so. The shoulder region is absolutely massive in all dimensions. Viewed from above, the bull has a wedge-shaped body, but shaped in the opposite way to the cow. In an ideal dairy bull, the rump is narrow and the shoulders are wide. The massive shoulder area extends into a slight hump along the spine, located between the shoulder blades, in the region of the cowlick.

Bulls should appear as totems of strength and power. Their proportionally blunt and thick form is an indication of their physical dominance. The more developed the sexual characteristics of our bulls, the better female dairy offspring they will sire.

Hair Patterns

Astute farmers over the centuries have discovered that external patterns of hair growth give tremendous insight into the internal health and vitality of our cattle. Hair patterns give strong universal indications regarding the functioning of the endocrine, immune, and reproductive systems.

At the most basic level, we want our cattle to have sleek and shiny hair coats. The quality of the animal's coat indicates its state of nutritional balance. Well-nourished and fully mineralized animals will have healthy immune systems. A thriving animal naturally sheds any dirt or debris from their coat, keeping them clean and healthy. We cannot directly see the cow's internal mineralization, but we can observe a sleek and shiny coat as an outward indication of internal health.

It follows that animals with bristly coats, frizzy hair, and a matted or dirty exterior have something compromised in their internal condition. Calves suffering from any sort of parasite problem are particularly obvious in this regard. We generally do not need to do a stool analysis to recognize that a calf is suffering from an unhealthy parasite burden. These calves will tend to have manure stuck to their tails, and their hair will appear wiry and stiff, almost like they are being jolted by static electricity. Calves with this appearance should be easily identified as unhealthy animals, and treated appropriately.

Moving down the animal, from head to toe, let us now consider the many subtle insights that the specific patterns of hair growth demonstrate to the observant dairy farmer.

All animals should have a spiral swirl right in the middle of their face. This spiral, located between the animal's eyes, tells us a lot about the animal's disposition and temperament. The two things to examine closely are the quality of the spiral, and its exact location. The spiral should be clear and symmetrical. Broken spirals or irregular forms indicate some form of mental derangement in the animal. The location of the spiral indicates the docility of the animal. A spiral that is centered above the line between the cow's eyes indicates an animal that is inclined towards an ill temperament. The facial spiral should be located beneath a line drawn between the eyes, indicating a balanced mental constitution.

The tuft of hair between the cow or bull's horns is a distinct growth of hair. In a robust and healthy animal, this hair should be thick and tight, appearing like a nice set of 'bangs'. This tuft of hair should be distinct in character, noticeably different that the surrounding hair of the head. Neat and trim, it should not appear overgrown. It may be a different color, but should be comprised of healthy and luxurious hair.

The hair growing on top of the head gives strong indications regarding the fertility of our bulls. The hair on the bull's head should be curly, and the tighter the curls, the more fertile the animal. The curls should spiral tight to the skull. The curly hair of the bull's head will ideally extend down the bull's neck, giving the entire head region a powerful and striking appearance. Waviness in the hair, or even worse, hairs that are sticking straight up, are signs of compromised fertility in the bull. Any time an animal's hair looks like it is being electrocuted, this is a very bad sign for its fertility and underlying genetic quality. A detailed discussion, complete with comparative photographs demonstrating varying states of fertility, can be found in James Drayson's book, *Herd Bull Fertility*.

On the neck of the all bovines, originating from the groove of the jugular vein, the animal should display a 'thymic swirl'. The hair patterns in this area of the neck, corresponding to the animal's thymus gland, are indicative of the health of the immune system. Whereas the general hair of the neck lays downward, the thymic swirl should be an area of finer hair, growing upward in direction. The thymic swirl is present on both cows and bulls. A good thymic swirl indicates a healthy thymus gland, essential for a strong immune system.

Atop the bovine's spine, generally near the shoulder blades, is their cowlick. The cowlick is a small area of hair that grows in the opposite direction of the rest of the hair along the spine. The further forward the location of the cowlick, the better the animal's reproductive health. The location of the cowlick should never be to the rear of the shoulder blades. The hair of the cowlick is indicative of the cow's milk quality, and is also strongly correlated with her estrous cycle. The hair of the cowlick will typically stand straight up when a cow is in heat. Occasionally, an animal will have two cowlicks, also called a 'double cowlick'. This is a positive indication of reproductive health and overall fertility in both cows and bulls.

The base of the bull's tail is an area that indicates his testosterone levels. The bull should have thick and curly hair atop the head of his tail. The appearance should be similar to what we want to see on top of the bull's head. In the cow, these characteristics would be undesirable, as they are indications of masculinity.

The tails and flanks of our cows and bulls should be clean and shiny. A tendency to be dirty or to accumulate manure is an unhealthy indication. With a strong immune system and good circulation, the tail should remain clean naturally. When we see animals in a feedlot, covered with manure and mud, it is a bit of a chicken-and-egg situation. The animal is not dirty only because of its filthy environment. The animal is dirty because the filthy environment has compromised their immune system, so they lose their natural ability to keep themselves clean. Healthy environments lead to healthy animals that stay clean and shiny naturally.

The Escutcheon

The escutcheon is the area on both cows and bulls that is located between their nipples and their anus. Broadly speaking, this is the greater genital region. The form of the escutcheon is the most profound indicator of dairy quality in our animals. The physical characteristics of the escutcheon are fixed at birth, and do not change at all through life. The escutcheon can be thought of as a fingerprint, unique to each animal, and highly indicative of dairy performance.

In the nineteenth century, the French farmer M. Guenon compiled his observations of escutcheon patterns and dairy production. In 1888, his landmark text was translated into English, titled *Guenon on Milch Cows*. The insights of Guenon have been confirmed by modern dairy farmers, including holistic veterinarian Paul Dettloff of Minnesota. In this chapter, these findings are significantly consolidated and simplified for readers of this book. For the serious student of dairy farming, obtaining a reprinted copy of Guenon's work, complete with dozens of original detailed illustrations, is highly recommended.

The specifics of escutcheon forms are complex, though readily understood with the aid of Guenon's extensive illustrations. Fortunately for modern dairy farmers, *Guenon on Milch Cows* has been recently republished and is easily available. The usefulness of this text cannot be overstated for dairy farmers seeking to improve the milking quality of their herd.

The fixed character of the escutcheon allows the dairy farmer to evaluate and compare the milking qualities of his calves in the first month of their lives. Typically when a calf is newborn, the hair growth on the escutcheon will not be developed fully, and evaluation will be difficult. By a month of age, however, the escutcheon patterns should be clearly visible, and will provide great insight into a given animal's future value. There are no quantifiable absolutes to escutcheon characteristics, but we can gain significant insight into our animals' future dairy quality without waiting three years for them to become mature milk cows themselves.

The hair growth that covers the genitals and the escutcheon of both the cow and the bull is distinct in character. It is downy and soft, very fine, and almost silky to the texture. In contrast to the common hair that grows downward, the hair of the escutcheon grows upward, like the hair in the thymic swirl. The extent to this downy escutcheon hair indicates both milk yield and milk quality, in both the cow and the future female offspring of our bull.

In the simplest terms, we are looking for the largest escutcheon area as an indication of greater udder capacity and higher butterfat content. This is true of both the cow and the bull, though the escutcheon is comparatively much smaller in the bull than in the cow. The bull obviously does not produce milk himself, but the form of his escutcheon indicates the dairy qualities that he will pass on to his female offspring.

The Cow's Reproductive Area

The cow's vulva should be large and swollen, indicating a strong flow of blood to her reproductive organs. Any hair on the vulva should be fine and soft, like the hair on her escutcheon. A hairy vulva is an indication of hormonal derangement and poor fertility.

The area between the udder and the vagina is the location of the cow's escutcheon. The escutcheon will cover the rear part of the udder, and extend upward toward the vagina. As discussed in detail in Guenon's text, there are different basic 'forms' of escutcheons, and the escutcheon will not necessarily extend upward all the way to the vagina. Comparing different escutcheon forms is a bit of an apples to oranges comparison, as a 'curveline type' escutcheon will always be shorter than a 'flandrine type' escutcheon, though both types can equally indicate excellent milk production.

In top producing milk cows, the escutcheon will be broad, with the area of fine hair extending outward onto the animal's thighs. Wide escutcheons indicate persistence of milk production through the lactation. Another important consideration is that there are no disruptions within the escutcheon. Islands of downward growing hairs, called 'feathers' by Guenon, that are located within the area of the escutcheon, are generally indications of derangement, although there are exceptions that are described in detail in Guenon's text.

The Bull's Reproductive Area

The characteristics that we desire for our bulls' genital region are generally similar to those for our cows. The testes of the bull should be massive in size, and covered with fine hairs. Coarseness of hair, or uneven size in the testes, are both very poor indications for fertility.

The escutcheon area on the bull begins on the scrotum, and extends upward. It is generally much shorter in total height than on the cow, only rarely extending all the way to the anus. The bull escutcheon should still be very wide at its base, expanding outward onto the animal's thighs. The same desire for fineness of hair, without any interruption or distortions, apply for bull escutcheons. There are fewer escutcheon forms for bulls than for cows, so comparing our bulls is more straightforward than comparing cows.

Examination of the escutcheon is a profoundly helpful tool when deciding which animals to sell and which to keep. This applies significantly to the bulls that we select for breeding. Escutcheon evaluation is our most powerful tool for understanding the milking quality that a bull will pass on to his offspring. Size and quality of bull escutcheon should be our most important factor to consider when selecting our breeding bulls.

8. HOLISTIC HERD MANAGEMENT

Holistic management of the small dairy cowherd rests upon three fundamental principles.

- Efficient farm labor so that the farmer enjoys their commitment to dairy farming.
- Feeding a natural diet for the health of our cows and their milk.
- Harmony with the seasons to create a life rhythm that works in concert with nature.

These three principles form a stable platform for running a small, holistically managed dairy herd.

The Farmer is the Most Important Animal on the Farm

Dairy farming involves a greater degree of stewardship than any other form of agriculture. The needs of the dairy herd are unrelenting and require frequent attention. It follows that the daily care and management of a herd of milk cows must be handled efficiently. The potential for exhaustion and frustration must not be underestimated. The daily milking routine is a highly skilled task, which cannot easily be hired out to ease the labor burden. As such, developing a system of management that focuses on the farmer is essential.

The single largest responsibility faced by the dairy farmer is the daily milking of the herd. Each cow will take 20-30 minutes for milking, plus another hour total for setup and cleanup. The size of our herd ultimately will be limited by many factors, including land area, market potential, and farm labor. The single biggest management choice that the farmer can work to his advantage is a shift from the traditional twice a day milking, to a holistic, once a day milking routine.

The historical pattern of twice a day milking was born out of industrial farming, where the labor was provided by hired hands. Industrial farming further pushed the necessity of twice a day milking by feeding cows grain-based concentrates, unnaturally forcing a dramatic increase in milk production. In a holistically managed system, once a day milking works perfectly, and results in a much more agreeable labor situation for the farmer. So many potential dairymen are dissuaded from starting a milk herd because of the unreasonable burden of twice a day milking. Small farmers of today realize that regularly hiring skilled outside labor is simply not financially viable. It is too much for one farm family to be responsible for milking twice a day, in addition to all the other chores on a diversified family farm.

Once a day milking can be scheduled at whichever hour of the day is most convenient for the farmer. So long as the same hour is observed each day, the cows do not mind being milked at sunrise, mid-morning, late afternoon, or early evening. With twice a day milking, the farmer must time their schedule to be twelve hours apart, which immediately creates a real math problem. This difficult reality is the reason that dairy farmers are assumed to be awake before dawn, so that they could milk at 6am and 6pm, finishing their evening milking by 9pm. Over the long term, this is an untenable schedule, which fails to respect that the farmer is the most important animal on the farm. Once a day milking allows the farmer to comfortably schedule his day; milking at a time that is practical for his family, and harmonious with the broader life that he lives.

Many a conventional farmer is shocked by the idea of once a day milking. Having no practice with such a system, they tragically dismiss something that could bring a dairyman such ease in their farming life. Once a day milking does yield less milk per day than twice a day milking; that is true. There is probably a total production loss of 30% from once a day milking versus twice a day milking. Fortunately, for this 30% drop, there is a full 50% labor savings. That math looks pretty good. Additionally, the extra 30% milk from twice a day milking results in a significant increase in the nutrient needs of the cow. Once a day milking keeps our cows in better health and condition, as their nutrient needs are more easily met on pasture with a slightly lower daily milk production.

Once a day milking is more reasonable for the farmer, and healthier for the dairy cow. It is an essential management decision that makes small-scale dairy farming more sustainable for all involved.

Natural Diet for Healthy Cows and Nutritious Milk

The nutritional requirements for dairy cows being milked once a day can be fulfilled by a diet comprised exclusively of pasture forages. There is absolutely no need to feed our cows any grain or unnatural concentrates. The bovine digestive system, with its extraordinary rumen and four stomachs, was perfectly designed to extract nutrition out of grass and other plant forages. Feeding a cow grain is analogous to putting rocket fuel into a motorcycle. Using the wrong fuel causes rapid mechanical failure. It follows that the majority of health problems in conventional dairies are directly caused by the feeding of an unnatural, grain-based diet. The natural diet of pasture forage is a key component in the holistic management of healthy dairy cows.

The cow's complex digestive system is a true wonder of natural design. Utilizing a series of individual fermentation vessels, or stomachs, the cow uses microorganisms to break down plant material into bovine nutrition. You do not actually feed the cow. You feed the bacteria in her rumen, and she lives off the excrement from this bacteria. It is critical to consider what the cow's gut bacteria are designed to eat themselves, as their health will determine the wellbeing of our dairy cow.

When cows eat the cellulose-based feeds that their biological system evolved with, they enjoy excellent health. When unnatural foods are introduced, their system is not evolutionarily adapted, and digestive distress results. Grain does not nourish the indigenous microorganisms of the cow's digestive system, and their healthy populations decline. Unnatural bacteria flourish in the vacant environment, which causes as shift in the pH of the cow's entire body, to a more acidic condition. Acidosis, as it is called, causes a weakening of the immune system, and opens the doors to all sorts of systemic health problems.

Acidosis from grain feeding is the number one cause of mastitis. With an acidic blood condition, the cow's immune system cannot work optimally to prevent bacterial infection in the udder. Even seemingly minor, subclinical mastitis outbreaks cause an increase in the bacterial content of the raw milk, measured scientifically as the 'somatic cell count'. As producers of raw milk, we will not be utilizing pasteurization to kill these bacteria; therefore, it is critical that we manage our dairy cows for optimal udder health. Elevated somatic cell counts are simply not acceptable in the raw milk dairy. Our milk is healthier, tastes better, and is more nutritious when we feed our cows a natural diet of pasture-based forage.

Cows that consume a natural diet, based exclusively on pasture forages, produce superior raw milk. Their milk has higher levels Omega-3 fatty acids, and Conjugated Linoleic Acid (CLA), both super-nutrients for human health. Pasture forages have higher mineral levels than grain-based concentrates, resulting in healthier cows and more mineral-rich milk. The finest cheeses from Switzerland come from cows grazing exclusively on high mountain meadows. The delicious flavor of these cheeses comes directly from the quality pasture forages that the cows were consuming during their lactation.

Sprouted grains have been touted as an alternative to conventional ground concentrates, purportedly with better health effects for the cow. In reality, sprouted grains are still an unnatural food, with unnaturally high starch levels that disrupt the proper balance of bacteria in the cow's digestive system. Cows did not evolve to eat sprouted seeds in any quantity, so the microbes in their digestive systems do not handle them optimally. Acidosis will still compromise the health of cows eating sprouted grain.

It is true that in nature, cows do graze mature seed heads on pasture plants. This is perfectly natural, and the amount of actual grain consumed by the cow is proportionally offset by the amount of fibrous husk and stem that comes with the seed head. Pasture fed cows will consume small amounts of mature grass seeds; this is not a problem.

There are many factors to consider when managing the cow's diet on pasture-based forages. An entire chapter will be dedicated to this critically important topic. For now, it is important to simply understand that holistic dairy cow management involves no grain in the diet. Our cows are healthier, and their milk is more nutritious, with a forage-based diet based on the natural biology of the cow.

Working with the Seasons

All relatives of the cow give birth to their offspring during the lush growth of spring. Deer, elk, bison, antelope, buffalo, and moose all have natural birth cycles that work in harmony with nature. As dairy farmers, we manage the breeding cycles of our cows, and seek to imitate nature as a means to greater health for our herd. Cows are herd animals, and for optimal health, we manage our individual dairy cows as one holistic herd unit.

Calves are born as spring is just arriving in full. Calving occurs naturally, out in a fresh grass pasture, typically with no human intervention. There is no more sanitary or healthy place for a cow to give birth. The dangerous cold of winter is past, and there is no need to provide shelter to the newborn calf born in spring. The last light frosts of the year will not harm a healthy young calf, so we should aim to see our first calves born just before our traditional frost-free date. This timing gives the momma cows a good spring tonic period on new grass, and also maximizes the productive length of our milking season.

The entire cowherd should be bred at the same time in summer. In doing so, calving becomes a distinct annual event in the life of the herd. Our cows benefit from the maternal support they give one another. In their amazing bovine way, cows coach each through calving, and babysit each other's calves in the summertime. Our milking chores are consolidated into a distinct milking season, so that we gain the economy of scale from milking all of our cows at once. The cows are all bred simultaneously at the height of summer, when their fertility is at its natural peak. Finally, at the start of winter, all of our cows are dried up at the same time, and both cows and farmer earn a deserved break during the dormant season of winter.

The first lush growth of grass in the spring is a medicinal tonic to our cows. After a winter of eating dried forages and hay, the shift to succulent feed in early spring has significant health effects. Fresh, tender dandelion leaves and grass tips are detoxifying and laxative for the cow. We time our cows' calving so that the cow has a few weeks of grazing these fresh forages prior to giving birth to her young. Fresh plant leaves cleanse the cow's blood, in the healthful tradition of spring tonic salads. The laxative effect of lush plant material actually makes calving easier for the cow. The first fresh pasture of the year helps to prepare the cow's system for birth.

Pasture is our primary currency as holistic dairy farmers. As such, our milk production closely parallels the growth curve of our pastures. We manage our herd to milk our cows when the pasture provides adequate feed for their lactation needs. In a fertile, North American climate, we can expect a natural milking season of 6-8 months. On our farm in Zone 6 Colorado, we typically milk from mid-April to Thanksgiving. By Thanksgiving, our pastures do not have the nutrient levels necessary to sustain high levels of milk production. The cows' natural lactation curve is declining significantly by this time, independent of the pasture quality. There is still forage available in the pasture; it simply is not dairy quality anymore.

The cows are harmoniously dried up at this time, as winter begins to set in on the farm. The remaining low quality forage in the pasture is perfect for a drying up the cows. Low quality roughage is the ideal food for a cow that we want to stop milking. We make use of the final month of grazing, well into December or beyond, to dry our cows up. Assuming that our cows have calved naturally in the early spring, and were bred at the peak of summer, this time in early winter will be their period of lowest nutritional needs of the entire year. The seasonal conditions of the environment are in perfect harmony with the natural nutritional needs of the cow. We are farming in harmony with nature, and economic and health benefits come naturally.

The final phase of the year is deep winter, when our cows are growing their young in their bellies. We will have exhausted the remaining forage in our pastures earlier in the winter. The farm is now frozen and dormant, asleep in the depths of winter. This is our hay-feeding season. Our cows need the good nutrition of alfalfa or clover hay to grow their calves optimally. In the final two months of gestation, the cow's nutritional requirements increase dramatically as their calf nears term. This is the ideal time to feed top quality hay. Working with the natural rhythm of the seasons, we have synchronized the cow's annual cycle in a way that benefits both the cow and the farm.

> *Every deviation we make from the natural annual cycle of the cow costs us significantly. The economics of running a small dairy herd are dependent on us making optimal use of the natural resources of the farm. Timing, in farming and in life, is everything. We are not running a factory, we are stewarding a natural system of agriculture. Nature is our ultimate boss, and we must cooperate with Her established rhythms. Operating in concert with nature is a satisfying and profitable way to manage the small farm dairy.*

Management through the Year

An idyllic image comes to the mind of a natural bovine herd, with young and old alike grazing through a fertile meadow. In reality, differing management goals will require us to separate various animals into subgroups. The most obvious breakup of the herd occurs at weaning time, when the calves are large enough to be independent of their mothers. Another breakup of the group occurs as we manage the breeding cycle of the cows, by removing the bull to prevent an unwanted early pregnancy. What follows is an annual description of the management of the individuals within the ever-shifting cowherd.

Out to Pasture in the Spring

Winter comes to an end with our pregnant cows together with any bulls in the hay yard. Our open heifers will be separate in the calf yard, securely away from any bulls. Preventing our bulls from breeding our heifers in the early spring in essential. Otherwise we would be setup for a disaster, with these heifers then due to calve early in the winter.

The moment when we move our herd out to pasture for the spring is the ideal time to harvest our yearling bulls for beef. The specifics of bull beef will be discussed in detail later in this chapter. Timing our beef harvest to coincide with our annual move to spring pasture will greatly simplify our herd management. Our cows can then head out to pasture with just our breeding bull, who will watch over the cows as their pregnancies develop.

At that time, any open heifers can either be grazed in a separate pasture, or if there are only one or two heifers, it may be preferable to tether them on a rope. The yearling heifers are still small enough that they are easily controlled by a rope tied to a halter. The daily interactions with the heifer on a halter helps to socialize the young animal, and make them associate the farmer with good things, in this case, fresh grass.

A fifteen foot rope attached to a their halter, allows them access to one day's worth of grazing at a time. They will need a water tub, and will drink roughly five gallons of water daily. Using a carabineer attached to the end of the grazing rope enables easy moves of their grazing location by simply wrapping the rope around a fixed location, like a post or a tree, and clipping the rope to itself. Cattle are docile enough in disposition to calmly respond to grazing on a tether, a common system in times past when grazing resources were more intensively managed.

Calving Time

The cows will calve out in the pasture, watched over and protected by the herd bull. As each cow delivers their calf, they will walk off alone, and birth in whatever relative solitude they can find. This is all in perfect harmony with the natural instincts of the cow.

For the first week after giving birth, the baby calf should remain with the momma cow at all times. During milking, the baby calf can wait outside the barn or in an individual barn stall, and then be immediately reunited after the cow is milked. The first week of life is especially critical for the calf, and we want the maximum amount of nurture and care from the mother at this time.

Once the calf is one week old, we will begin separating the cow and calf at night. Of course, this assumes a morning milking time. If milking were to occur in the evening, then we would want to separate them first thing in the morning. The idea is to allow 10+ hours of separation before milking so that the momma cow gets a full udder of milk.

The cow and calf pair is initially brought into the barnyard, and then the calves are separated off into the calf yard. Sometimes this goes smoother than others. Practice and patience will get everybody on the same page. One additional challenge is that our cows will not all be calving on exactly the same day, so we will have cow/calf pairs in different stages of management at any given time. Fortunately, cows and calves both learn from their peers, and once you have several calves that know the routine, it will be easier for others to learn when it is their time to be separated. Persistent patience in the proper position is the mantra of the day.

After milking, the calf is promptly returned to the momma cow. The calf then has the remaining half of the day (or night, as it may be) to nurse freely from its mother. The rhythm of half a day nursing works out perfectly for everyone. The farmer gets an ample amount of milk during milking. The cow gets her udder butted and massaged by the calf. The calf gets more than enough milk to promote healthy and vigorous growth.

These benefits for the calf, and the cow's udder, are other important virtues of the once a day milking routine.

Separating the Bull

Cows will experience their first heat after calving as soon as three weeks after giving birth. Before this heat period arrives, the breeding bull must be removed from the cowherd, to prevent an undesirably early calving date the following season.

The bull will need to be removed from the cowherd for roughly two to three months, depending on the exact breeding calendar that is being followed. The specific details of these considerations and calculations will be discussed in Chapter 11 on Breeding and Calving. For now, the bull simply needs to be separated from the cows. He cannot comingle the open heifers, as they need to remain unbred at this time as well. The bull needs to be isolated for two to three months to maintain proper timing in the breeding cycle of the herd.

The bull can be grazed on a tether, attached to his nose ring. If this is desired, to simplify management, the open heifers can be combined in with the newly lactating momma cows. This way, there is only one animal on a tether to handle. Grazing a bull on a tether is straightforward, and operates exactly the same as grazing the heifers. The only additional consideration is the safety awareness required when handling a large and powerful dairy bull. This too will be discussed in detail in Chapter 11 on Breeding and Calving.

The other option for isolating the bull involves the construction of a dedicated bull shed for containing the bull. The bull shed needs to be constructed to a very high level of sturdiness. Posts should be six inches in diameter, and anchored deeply into the ground. The boards that enclose the sides of the shed should be two inches thick, and secured to the inside of the posts with screws, for maximum strength. In addition to the four corner posts, a line post should be located along the midpoint of each wall section. Never underestimate the power of a dairy bull, and never give him a chance to know his own destructive power, especially when there is a nearby heifer to be bred.

A traditional bull shed should be at least fourteen feet square. It will need a roof, and excellent ventilation around the tops of the walls. The walls should remain solid up to six feet in height, followed by a vent area of about eighteen inches. The doorway to the bull shed must be constructed very solidly. A small hay door, located conveniently for feeding hay, is necessary. The hay door should be large enough to comfortably feed hay, but small enough that the bull cannot attempt to rush through this opening during feeding.

Two rubber buckets should be securely anchored to the wall, immediately adjacent to the hay door. This way, the water buckets cannot be pushed around the bull shed. Under no circumstances should the farmer ever enter the bull shed when it is inhabited by the bull. The dangers are simply too severe to risk.

Each day, hay is fed to the bull, generally thirty pounds at a minimum. The water buckets are filled, and a salt block is provided. There is a significant cost to having to supply hay to the bull for a two or three month period, but sometimes, isolating the bull in the bull shed is the simplest and most holistically sensible strategy for bull management.

Having a proper bull shed is a worthwhile investment for the dairy farmer, so that in any emergency, the bull can be secured with complete confidence.

Weaning the Calves

Calves continue nursing for half of each day until they are two or three months old. By now, you should notice the calves actively grazing out in the pasture on a regular basis. Their nursing will have become quite aggressive, to the point that you wonder if it is helping or hurting the cow's udder. The calves should appear plump and stocky. Weaning the calves is a simple process of not returning them to their mothers one day.

The best way to wean your calves is to wean them all at the same time, despite being some weeks apart in age. Wait until the youngest calf looks amply vigorous; you do not want to overly short change your youngest. Weaning is a noisy time. The calves will bawl and bawl, often bawling themselves hoarse in the process. The momma will moo and moo, sounding very distressed. This is a normal, albeit painful part of the weaning process. Do not vacillate and return the calves to their mommas once the weaning process has begun. Doing so will do no favors to anybody.

At weaning time, the calves should be moved to a small area of excellent quality pasture. By the time the calves are weaned, it should be high summer, and an area with good clover should be their first destination. They calves will not eat a lot at first, so a small area will suffice. But that area absolutely must contain the very best quality forage to make up for the sudden loss of milk nutrition in the calves' diet. The weaned calves need fresh water and a salt lick in their pasture as well. Ideally, you should locate the calf pasture such that the calves can see their momma cows out in the field grazing. Visual contact between cow and calf will help to ease the stress of weaning.

Breeding Season

Almost immediately after the calves have been weaned, it will be time for breeding the cows. Depending on the composition of your herd, if there are yearling heifers that you are not going to breed this year, they can be sorted out, and grazed together with the weaned calves. A yearling heifer will act as a big sister for the young calves, furthering their education in grazing, and helping to keep the young calves groomed. Like a teenager, such a yearling heifer may somewhat resent not getting to be with the grownups, and still being stuck with the kids. But since she cannot be allowed near the bull this year, keeping your yearling heifers with the young calves is the best management for your herd.

The exact time that you want to introduce the bull to your cows will depend on your breeding plan. This will be discussed in detail in the chapter on breeding and calving. For now, roughly speaking, we can assume that sometime in mid-summer, you are going to want to put your bull in with your cows. At this point in the season, you will have your calves and yearling heifers together in the calf pasture, your bull on a tether, and your cows and second year heifers together in your main pasture.

Bringing the bull in with the cows will generate a lot of excitement in the herd. Your bull will be very excited by the reunion with his cows, and you need to be careful at all times. Ideally, your bull will be so infatuated with his harem of cows that he will pay you no attention at all. However, when you are herding the cows into the barn for milking, he may not be happy about you taking his cows away from him. Be aware that in nature, the only thing that could take away a bull's cows is another superior bull. Your bull may view you as competition for his harem, and may feel the need to challenge your superiority. Needless to say, this would create a very dangerous situation.

Herding the cows into the barn for milking, you want to train your bull to wait outside. You do not want a large and potentially destructive bull inside the confined area of the barn at any time. If your bull insists on following the cows into the barn, make sure that his time in the barn is the worst experience of his life. You can throw things at him, shout, or strike him with a poker, anything to make him associate the inside of the barn as a truly horrible place for a bull to be. He will learn quickly, and will walk to the barn from the pasture with his cows, and then turn away as the cows walk into the barn. You will have previously trained your cows to come into the barn, so that should be no issue, they will continue about their routine, independent of the bull. Training the bull not to enter the barn is generally an easy lesson, which should happen with just a few days of consistent effort.

The Smooth Days of Fall

The management of your herd can stay the same in the fall as it was at the end of summer. So long as your bull is not being disruptive, he can stay with the cows in the primary pasture. Your calves and yearling heifers will stay together in the secondary calf pasture. A stable routine is best for your animals, who should be perfectly comfortable with their situations by this time.

Easing into Winter

As winter approaches, seasonal shifts in management will depend primarily upon the nutritional content of your pasture. Secondarily, severe winter weather will begin to affect your young calves.

The calves and yearling heifers will be your first animals to need increased shelter and access to hay. Our mature animals can graze through the snow, but we cannot expect our calves to do well with this expectation. As soon as the first significant lasting snows of the year arrive, bring your calves and yearling heifers into the calf yard, and begin feeding them hay for the winter.

As your pasture begins waning in forage quality, if you want to extend your milking season for a few more weeks, you could begin feeding hay to your dairy cows a little early, as a means to extending their milking season. If you decide to do this, leave your bull and bred heifers out at pasture, as this will economize on your purchased feed costs. Just bring in your lactating dairy cows to the barnyard, and feed them hay exclusively. Depending on the amount of hay you have, and the value of continuing milk production for an additional few weeks, this can be a good solution to extend your milking season into the beginning of winter.

Eventually, cold temperatures and low production will persuade you to stop milking your cows for the year. This is a bittersweet day for the seasonal dairy farmer. Fresh milk is over, sales come to a halt; but all full day off is yours! Return your dairy cows to the pasture, reuniting them with the bull and bred heifers. The remaining scant bits of low quality forage will be perfect for drying off your cows. You cannot dry off your cows while feeding them dairy quality hay. Many times, especially if you have some woodland area for them to shelter in and browse upon, you can keep these animals out at pasture well in January. This reprieve from hay feeding will result in significant economic savings for the farm.

Monitor your cows during this period. It is okay for them to be a little hungry. This is a natural part of their evolved annual nutrition cycle. We certainly do not want our cows to be starving, however. Look at the barrel of their rib cage; if this starts to appear sunken, it is past time to bring them in for hay feeding. If truly severe winter weather sets in, especially settled snow depths deeper than one foot, or ice storms of any type, this will end the grazing for the season. Every year is different, and your farmer's eye will need to monitor the land and your cows carefully. Do try and wait to begin feeding hay, as once you start feeding hay for the winter, you will be committed until spring. The cost of feeding hay is the most substantial annual expense for the small dairy farmer.

Hay Season

Finally, all your animals will be brought into either the cow yard or the calf yard, and will be living off of your stored hay.

The final consideration you will have is that at some time early on in the winter, your bull calves will become fertile. It is almost unbelievable, but vigorous dairy bull calves are capable of breeding at nine months of age. In order to prevent your heifer calves and yearling heifers from getting bred prematurely, sometime in mid-winter, shift your bull calves from the calf yard to the cow yard. Your cows will all be pregnant, so the bull calves cannot have any effect upon them. Make sure that your cows udders have been completely dry and empty of milk for some time, so there is no danger of the bull calves nursing when reunited. If the bull calves were to start nursing their moms before they are fully dried up, the cow would be stimulated to continue producing milk, and this would place an undesirable strain on her physical health during winter.

At the end of the winter, you will have your bred cows, bred heifers, and bulls all together in the main cow yard. Your heifer calves and open yearling heifers will be together in the calf yard. Hay will be the exclusive feed for the herd, and fresh water and mineral salt will need to be provided.

The final task at the end of winter is to butcher our yearling bulls for beef. While the weather is still cool, and before we begin the annual reshuffle of animals going out to pasture, this is the ideal time to harvest beef from our young bulls.

Bull Beef

An inevitable product of our annual breeding and calving is extra bull calves. Of course, every dairy farmer wishes and hopes for heifers, but the reality is that on average, half of your calves will be bulls. When we apply rigorous selection criteria to any potential bull to be used for breeding, we find that probably only one in ten bull calves will be good enough for breeding our herd. Of course, another farmer may not have quite as strict quality standards, and we may find a market for a few of our reject bulls. Ultimately, there will still be a lot of bull calves with no future in breeding, and raising them as bull beef is an excellent choice for the small farmer.

Bull beef, harvested at a young age, has excellent eating quality. The conventional beef industry has convinced people that the only good beef comes from steers, but this is completely false and self-serving. Steer beef is the norm today because feedlots could not handle hundreds of young bulls in confinement without some serious bovine rioting. Raised freely on pasture, on small farms with adequate space, young bulls are perfectly agreeable in temperament.

Young bulls grow much faster than a comparable steer. Their elevated testosterone levels cause increased vigor and faster weight gain. By raising our male calves intact, we additionally save on the cost of veterinary castration. There is absolutely no reason to castrate our bull calves, unless we are raising them to be oxen. There is a widely held belief that somehow bull beef tastes unpleasant, but harvested at eighteen months or less, this is not at all true. The beef from young bulls is deeper in color and richer in flavor, in fact. For those of us who believe in the saying, 'you are what you eat', then eating bull beef gives us the strength and vigor of a bull. Compared to steer beef, bull beef is superior meat in every way.

For ease of management on a dairy farm, as described above, bull beef is best harvested at one year of age. If you have an adequate separate pasture, well fenced from your heifers, you can profitably raise your bull beef until the following autumn, harvesting the beef at eighteen months of age. There is a substantial gain in yield from raising your bulls to eighteen months of age, though it may not be practical on your farm due to fencing and space constraints.

A young bull from a dual-purpose breed, harvested at one year of age, will yield approximately 150 pounds of butchered meat for the table. At eighteen months of age, the yield increases to around 250 pounds of cut and wrapped table meat. Bull beef can provide a significant amount of the homestead family's meat for the year. Any surpluses can be sold farm direct to your milk customers, who will frequently be eager buyers of such high quality homestead beef.

Poultry as Symbiotic Companions

Chickens and cows are the classic farm combination, and they make perfectly complimentary companions. There are many different systems for how to integrate chickens with your cows, but the reality is that it doesn't much matter. Your chickens will flock to the cows, like bees to honey. Chickens instinctually know that cows are purveyors of fertility, and that where there are cows, there will be plenty of bugs and nutritious grubs to browse. Cows graze down the pasture, giving chickens better access to the soil surface where all the good eating is located. You can rely on free-ranging chickens to find and follow your cows.

Cows benefit greatly from chickens, because chickens act as natural sanitizers in the pasture. Chickens will scratch and peck at the manure piles, destroying fly larvae and parasite eggs. Chickens will even cautiously approach a sleeping cow, carefully pecking at any bits of debris on the cow's coat. Poultry manure is high in phosphorous, an essential nutrient for cows. The poultry manure that falls on the pasture promotes rapid regrowth of the grazed forages. After calving, your chickens will quickly clean any afterbirth or blood that falls on the pasture. Chickens work tirelessly, cleaning up behind our cows, and further improving the fertility of our pastures.

9. PASTURE-BASED DAIRY NUTRITION

The feeding of our dairy cows is the most important management responsibility of the dairy farmer. By developing a diversity of seasonal pasture forages, our cows can instinctively harvest their own balanced nutritional needs.

There are many excellent books that address in great detail the subjects of rotational grazing management and pasture soil fertility. The two that are mandatory reading for all pasture-based cattlemen are *Management Intensive Grazing* by Jim Gerish, and *Quality Pasture* by Allan Nation. It should be noted that these books are primarily written by and for beef cattle ranchers. There are important differences with the management of a dairy herd. This chapter focuses on the aspects of grazing and pasture management that are particular to dairy farming.

Maximizing the Grazing Season

Length of grazing season represents a primary factor in the profitability of the dairy business. Certainly, there are fundamental climatic realities that influence a given location's productivity. However, pasture management is the single most significant factor that determines length of grazing season.

The ideal to strive for, in imitation of natural ecology, is a system of cow feeding that requires no stored feed. Eliminating or reducing hay dependency from the requirements of the cowherd will pay great dividends in the profitability of the farm. Farmers from Alberta to Alabama have successfully developed grazing management systems that successfully achieve that ideal, so the ability for cattle to feed themselves exclusively by pasture grazing is not geographically limited. Developing a localized system of cattle feeding that relies entirely upon grazed pasture nutrition is a highly desirable objective for every dairy farmer.

The reality is that some limited, seasonal hay feeding will likely be a sensible compromise for the needs of both farmer and dairy cow. Unexpected weather events such as ice storms, deep snows, or prolonged mud seasons, must be dealt with as they happen. A system that depends completely on pasture feeding does not have much versatility or built-in resiliency with which to address these challenging circumstances. Forcing our cows to continue grazing during brief periods of extreme rainfall, for example, can result in mud damage to our pastures that can take months to fully recover. Stored feeds form a safety mechanism that we can utilize to the holistic benefit of our farm.

Bringing in hay from outside the farm represents an opportunity to import different forages, with different mineral and nutrient profiles than our native pasture. Like in humans, diversity of diet prevents nutritional deficiencies. A brief period of alfalfa hay feeding provides abundant plant minerals to our cows, minerals that their bodies can store and utilize throughout the year.

Holistically managing the nutritional needs of our cows, the ecological needs of our pastures, and the financial needs of our farm, is the challenge for the pasture based dairy farmer. For this, and many other decisions on the farm, the framework developed by Alan Savory in his brilliant work, *Holistic Management,* is helpful beyond imagination.

Beyond "Grass-Fed"

Many farmers refer to their cattle operation as "grass-fed". For the progressive dairy farmer, grass farming is just the tip of the iceberg. There is an entire cast of diverse pasture species that provide better nutrition to our dairy cows than grass could ever produce. Grass is valued for its resiliency, and its length of grazing season. Plants such as red clover, forage chicory, grazing alfalfa, broadleaf plantain, and dandelion are the pasture species that we work to make the mainstays of our grazing program. Grasses are used as support species, utilized primarily for their productivity during the coldest seasons of the year.

Through the year, we work with ever-changing, basic partnerships of plants in our pastures. Generally, we aim to have a leaf and a flower being grazed at any given time. The composition of our pastures is a fluid and changing canvas, never rigid in its makeup. In spring we will have orchardgrass leaves and dandelion flowers. In summer, our pasture balance will shift to broadleaf plantain leaf and red clover flower. In some cases, we may have forage chicory leaf and alfalfa flower, or even timothy grass and white clover flower. Fall will see a shift back to the cooler season forages of diverse perennial grass species, without a prominent flower species. This is a gross oversimplification, and at any time we should be able to find many species growing together in our pastures. However, this vision of a shifting ecological canvas helps to direct our management of our pastures.

Early Spring Grazing

As the last snows are melting away in spring, the first species to begin actively growing in the cold soils and frosty nights will be our cold season perennial grasses. Orchardgrass, in particular, is the superior forage for the early season. Its deep root reserves provide the necessary energy to produce vigorous new growth early in the spring. Orchardgrass is very cold tolerant. Its growth is large, thick blades of grass, growing in dense clumps in the pasture. Unlike the fine blades of many grass species, orchardgrass is substantive. Orchardgrass grows a large amount of biomass, and as such, functions well as the primary forage for our cows in the early season.

> *In many guides to pasture management, it is taught to never graze our pasture too short. This is absolutely true, with one critical exception: orchardgrass. Orchardgrass provides the majority of the cows' pasture needs in the early spring, and through heavy grazing and its natural slowdown in the summer heat, it is gradually pushed to the background, allowing better suited species to dominate our summer pastures.*

Dandelion is another spring pasture plant that we want to encourage in our fields. Dandelion is a small plant with a short grazing season. The sight of fields of bright yellow dandelion blossoms signals a definite transition of the seasons. On a sunny spring day, we can watch our dairy cows sauntering through a pasture, carefully plucking dandelion blossom after dandelion blossom, relishing the medicinal value of this wonderful plant. Dandelion is a cleansing plant for the cow's digestion and immune system. Complementary in form to the fibrous roots of the orchardgrass, dandelion has a deep taproot that helps to bring up subsoil nutrients and build topsoil. Dandelion blossoms provide excellent pollen forage, helping our native bee populations to get off to a good start for the year. As quickly as it emerges, dandelion is grazed and goes dormant again, fading into the background of our pasture ecology.

As spring advances, clumps of orchardgrass will aggressively cover the pasture. Orchardgrass needs to be grazed hard and repeatedly, while the plant is still young and palatable. Fast grazing rotations through our pastures, so that the orchard grass gets grazed every ten days, prevent it from undesirably maturing into coarse leaves and impenetrably dense clumps.

Mature clumps of orchardgrass will be rejected by our cows, as this over-mature forage does not provide adequate nutrition for a lactating dairy cow. These impenetrably dense crown clumps will prevent the natural succession of pasture species as the summer progresses. Each ungrazed orchardgrass clump will become a pasture sacrifice zone, effectively shrinking the functional area of our pastures for the remainder of the season. Proper grazing management of orchardgrass is essential in the springtime.

> *Tender spring grass has a perfect nutrient balance for pregnant and calving cows. As our cows transition into their lactation period, our pastures transition into summertime dairy pastures. The harmony between our cow's nutritional needs and our pasture's provision is perfectly balanced.*

Summer Legume Pastures

Summertime is the season of clover in our pastures. Heavily grazing our orchardgrass has created openings in our pastures, from which perennial clovers will emerge. As soon as we pass our final frosts in late spring, young clover plants will begin to grow rapidly. Our climate will dictate which clover species are most productive on our farm. Red clover is our preferred choice. In actuality, red clover is the single best species for dairy grazing.

Red clover in our pasture needs an ample rest period after grazing. In general, during the long days of summer, a 30-day rotation will work well with red clover. When grazing red clover, we only want to graze the top half of the plant, leaving behind the lower half of the plant to provide photosynthetic area to fuel its regrowth. The red clover should have reached a state of complete maturity when we let our cows in to graze it. It should be in full bloom, with just a few seed heads beginning to develop and dry out. At this stage, we maximize the available biomass for our cows. The red clover plants themselves will have developed large root systems, helping to enrich our soils. Allowing adequate regrowth of red clover between grazings is absolutely critical to its continued productivity in the pasture.

Pasture research has found that the ideal environment for grazing productivity occurs right around the 38-degree latitude line. At 38 degrees latitude, the summers are long, but not excessively hot. Natural rainfall tends to be plentiful. The 38-degree latitude line runs through San Francisco, the Western Slope of Colorado, the Nebraska and Kansas prairies, the Ohio River Valley, and the Appalachia of Virginia. To some degree, altitude and local climate can push this zone of optimal pasture growth north or south a bit. Anywhere near this latitude, red clover should thrive in a fertile, holistically managed pasture.

In regions where red clover is not ecologically ideal, our other pasture legume options will be alfalfa and white clover. Alfalfa excels in hotter, more arid climates. Alfalfa has a huge taproot that is tremendous in its drought tolerance and ability to improve soils. Alfalfa will benefit from an even longer grazing rotation than red clover, closer to 50 days between grazings. Fortunately, alfalfa will produce a larger amount of biomass in this time, so that its total annual productivity is on par with red clover. The same guidelines of grazing only half the plant, and allowing it to grow to full maturity before grazing apply equally to alfalfa.

The biggest management concern with alfalfa, grazed fresh in the pasture, is its propensity to bloat in cattle. Alfalfa must never be grazed when wet, either from rains or heavy dew. Even more critically, alfalfa cannot be grazed immediately after a frost. Observing these restrictions, alfalfa is an excellent pasture plant in hot climates. Allowing our alfalfa to be in full bloom when we graze it is critical for both our pasture and our cows. At this stage of maturity, alfalfa is both higher in minerals and less prone to bloat, providing the best possible feed balance for our animals.

White clover will be our choice in cooler climates that are too cold for red clover to thrive productively. White clover is a much more diminutive plant, which is its primary drawback. Grown to its maximum height, white clover produces much less biomass. Fortunately, white clover grows quickly, and only needs approximately an 18-day rest between grazings to mature properly. White clover provides good quality pasture nutrition for our cows, but it needs to be grazed more frequently to compensate for its lower biomass yield. The smaller biomass of white clover also belies a smaller subterranean root mass, so white clover is not as effective at soil fertility improvement as red clover or alfalfa, and certainly not as drought tolerant. Nevertheless, in cooler climates, white clover will be the most important species in the summer dairy pasture.

With all leguminous plants, such as red and white clover, and alfalfa, there is a process of nitrogen fixation that occurs every time we graze the plant. As the top of the plant is removed through grazing, an equal area of root is sloughed off underground. In legumes, which have nitrogen-fixing bacteria attached to their roots, this results in nitrogen nodules being released into the soil. This boost of soil nitrogen then enables other pasture plants to grow more productively. As pasture farmers, desirable companion forage species are established to benefit from this increased soil fertility.

The best complementary plants for our legumes are plantain and chicory. Of course, there are many grass species that will coexist in our pastures, but these grasses are not as nutritious for our cows or fertility generating for our soils. Grasses are the basic reality in any pasture; and they are good, but not great. Grasses are the default vegetation in any pasture, due to their natural resiliency and precociousness. Our management will focus on the prosperity of more desirable species like plantain and chicory, while appreciating the many grass species for the ground cover and forage that they do provide.

Plantain is the perfect companion to red clover, thriving in the same pasture conditions. It likes a moderate climate, moist soil, and adequate calcium levels. Plantain yields well with the same 30-day rest cycle as red clover. Pound for pound, plantain is nutritionally superior to any perennial grass in your summertime pasture. Plantain also is a deep fibrous rooted plant, which helps to build soil humus levels. The combination of red clover sloughing nitrogen when grazed, and plantain sloughing carbon, creates an underground compost factory. Well-managed red clover and plantain pastures will naturally build their humus levels through the grazing season. Plantain will also grow well in partnership with white clover.

Chicory is the perfect companion to alfalfa. Like alfalfa, it has a deep taproot, which makes it exceptionally drought tolerant and a great transporter of subsoil minerals. Chicory handles heat effortlessly. A longer rest period is needed for chicory, so it combines harmoniously with alfalfa in a hot climate rotational grazing program. The forage quality of an alfalfa and chicory pasture is excellent for dairy cows.

In a red clover pasture, chicory does well sown along the pasture edges, where it can be managed with electric fencing to allow periodic, but not regular, grazing access. Planted along fence lines or hedgerows, chicory will enrich the soil and provide extra feed for our cows in marginal soils. Chicory is a medicinal plant in its own right, in addition to its excellent nutrient profile. Incorporating such a vegetative superfood into our pastures benefits all aspects of our grazing and pasture health.

Managing Fall Grazing

The first heavy freezes of the fall will cause a dramatic change in the species composition of our pastures. Our legume crops, clovers and alfalfa, will cease to regrow once interrupted by heavy frosts. The remaining standing plant material may still be grazed, though the cautious farmer will not allow access to frost-killed legumes for a week after sustaining damage, as a precaution against bloat. As our pastures shift away from the summertime legumes, the ever-present perennial grasses will come to the forefront of our pasture productivity.

In the fall, cooler temperatures allow our cool-season perennial grasses to produce dairy quality nutrition once again. In mid-summer, we manage our pastures to discourage these grasses, as warm temperatures cause their forage quality to decline below the levels necessary for optimal milk production. Thanks to their inherent resiliency, these grasses are still present in the pasture, and now in the fall, with the natural disappearance of the legumes, our grasses receive the sunlight they need to resume vigorous growth.

The shorter days of fall mean that our grasses will necessarily grow much slower in fall than they did during the longer days of spring. Our grazing rotation will need to slow down, giving as much rest between grazing as possible. In the fall, it is generally not possible to rest too long between grazing. Allowing our forage to grow to its maximum size, just short of allowing seed heads to develop, will result in optimal forage yields. The declining natural productivity of our fall pastures will nicely parallel our cow's seasonal reduction in milk production.

The stockpiling of fall grasses for winter grazing is one final consideration in this season. Any pasture areas that we can allow to grow to maturity, and remain ungrazed, will become incredibly valuable as winter arrives. Grass pastures will handle hard freezes and heavy snows, with the forage retaining a substantial amount of its nutritional quality. We can think of these pastures as 'standing hay'. Our cows can harvest these forages later in winter, reducing our need for purchased hay. Of course, the challenge is that our pastures are naturally declining in productivity during the fall. So reserving pasture for later grazing only makes our present productivity declines more challenging to handle. Nevertheless, it is an important concept that standing grass forage in the fall can be reserved for grazing in the winter months, if a surplus pasture area exists.

Dormant yet Productive Winter Forage

As discussed in the chapter on Holistic Herd Management, we will dry up our cows at the start of winter. The process of drying up our dairy cows will be greatly assisted by their grazing on low quality pasture remnants. Trying to dry up a dairy cow while feeding her good quality hay is like trying to stop a car while simultaneously stepping on the gas and the brake. We want to manage our cows in harmony with the seasons and their attendant nutritional needs. As such, allowing our cows to subsist on the low quality pasture remnants as they are drying up is good holistic management.

In early winter, before ice, snow, and mud become problematic for the integrity of our pasture sod, our cows are best left out to pasture. Any dried seed heads, overgrown clumps of pasture, and woody shrubs will be eaten with relish during this time. All the food sources in the pasture that our cows rejected when there was succulent red clover available suddenly become attractive. Having our cows graze down this remaining coarse vegetation will promote better pasture regrowth in the spring.

The early winter is a particularly good time to allow cows access to any hedgerows or deciduous woodland areas on the farm. Our cows can shelter safely among the trees and brush, while deriving their sustenance by eating tree buds and even fallen leaves. Conifer and other evergreen trees have zero feed value for our cows, and should be specifically excluded from any grazing plan.

Feeding hay should be delayed as late as possible, first maximizing use of available farm forages. Encouraging our cows to consume a diversity of plants is good for their health. Enabling our cows to subsist without costly hay is good for our farm finances. Seasonal access to mature hedgerows and woodlands is healthy for those ecosystems, working to thin them out and creating opportunities for new plant growth in the coming spring.

Rotational Grazing and Portable Electric Fence

Controlling our cows' movement across the pasture enables the farmer to practice management intensive grazing. The farmer can focus his cows on the sections of the pasture that are at the perfect stage of maturation. Permanent pasture divisions are nowhere near as precise as portable electric fence. Additionally, permanent pasture fencing is expensive to construct and maintain. Electric fence is the economical tool that dairy farmers use to maximize the health of their pastures and the performance of their dairy cows.

When fencing dairy cows, a single strand of electric fence twine is adequate to contain grazing animals. Baby calves will be small enough to slip under the fence, but their mommas will not be able to follow. As such, calves may wander under the fence line, but they will not go far from their mommas. A single strand electric fence is absolutely not adequate to keep a breeding bull away from fertile cows. In general, so long as the grazing is good where the cows are, they will not test the electric fence to get to the 'greener' grass on the other side.

Portable solar-powered fence chargers work well for single strand electric fence. Deeply pounded grounding rods are necessary for the proper functioning of the fence charger. It will work well to pound a few grounding rods at several convenient locations in the corners of our farm, so that the fence charger can be moved as needed to more practical locations. There are many fancy and expensive options available for electric fencing. They are not needed for cattle. A small solar charger, simple steel 'pigtail' portable fence posts, and the most economical electric fence twine will work great. Locating a few metal T-posts in key locations in the pasture will allow you to affix your fencing twine securely to these key points. Otherwise, portable metal pigtail posts are perfectly effective for managing long runs of electric fence twine.

The basic system of management intensive rotational grazing relies on two sections of electric fence, the front line and the back line. The front line will get moved each day, allowing our cows access to fresh pasture. The back line will get moved every few days, to prevent the cows from re-grazing pasture that is just starting to regrow.

There are two reasons why we do not worry about moving the backline every day. First is to economize the farm labor requirements. Second, leaving the backline in place gives our cows a bit more area to spread out and relax. The cows will almost always graze the fresh pasture adjacent to the frontline, but allowing them a bit of extra space for laying and chewing their cud reduces stress from crowding. The back line is moved forward every few days, initiating the regrowth period for that area of grazed pasture.

Deciding how large of a fresh area to give the cows each day is the principle challenge of management intensive grazing. The difficulty is that the size of this area is in constant flux, changing throughout the year, depending on the yield of the pasture and the quality of the species being grazed. There are no simple guidelines that govern how much pasture per cow per day will be needed. As such, grazing management is far more of an art than a science. Developing your 'grazing eye' will be an essential part of becoming a good pasture dairy farmer.

Our grazing rotation is designed to enable our cows to consume the pasture plants at their peak of nutrition and yield. Ideally, we want to graze our grasses just before they begin to flower. Our legumes are best grazed when they are in full bloom. Especially with grasses, allowing the plants to over-mature results in a huge decline in palatability and nutrition. The grass leaves become coarse and fibrous, and our cows will not want to eat them. If we force our cows to eat overly mature grass, their performance will suffer. If we leave the old grasses ungrazed, they will occupy valuable space in the pasture, which will be unproductive for the remainder of the grazing season.

In a proper rotation, we graze roughly the top half of the available plant material, and then move the cows on to the next area. We can think of this as 'skimming the cream' off of the pasture growth.

Perfectly grazing half of the pasture plant in a given rotation performs two key objectives. First, the top part of the plant has the most energy for our cows. Secondly, leaving the bottom half of the plant maintains a large enough leaf area to enable good photosynthesis and rapid plant regrowth. Careful observation of the pasture is key to ascertaining our success in these objectives.

If we do not allocate enough area to our cows, there will be too much grazing pressure on that area of the pasture. The cows will graze that pasture section down very short. They will not be able to harvest adequate pasture nutrition, and their milk yield will decline. Additionally, the pasture that has been grazed down excessively will regrow very slowly, reducing total pasture yield for the season.

Too little grazing pressure is just as problematic as too much. With inadequate grazing pressure, large mature grasses develop and become unpalatable. These overripe sections of pasture will not regain dairy quality grazing for the remainder of the season. The result is equivalent to shrinking our available pasture area. Poor grazing management in the early season will compromise productivity for the remainder of the year. Our watchful 'grazing eye' must be constantly evaluating our pastures for any excesses or deficiencies.

The rate of pasture growth varies dramatically through the year. Long days, warm temperatures, and ample rainfall will cause our pastures to grow rapidly. Shorter days, frosty nights, hot weather, and drought will cause our pastures to grow slowly. The rate of pasture growth will influence how long of a rest we need to plan between grazings of a given section of pasture. Adequate rest periods will vary from ten days to sixty days, as discussed earlier in the chapter. In all things with our grazing management, we should have ideals in mind that we aim for. The reality of management intensive grazing is always less than perfect. In time, with practice, we develop our 'grazing eye', and learn how to manage our pastures more productively.

One tool that we have for managing our pastures, that can really help to compensate for mistakes in timing, is our non-lactating animals. Remaining within the bounds of the specific management requirements that we discussed in the chapter on Holistic Herd Management, we can utilize our bulls and heifers to graze overgrown pastures that have lost their dairy quality. Our lactating dairy cows have the highest nutritional needs, and as such always need to be on excellent quality pasture. Animals that are not lactating can handle much more coarse forages without suffering a loss of performance. When we have an overgrown pasture, we can utilize the non-lactating animals in our herd to graze that low quality forage, and restore the pasture to balance.

Location of traffic paths is a key consideration in the use of electric fence. Pathways will become trampled and degraded by the heavy hooves of mature dairy cows, walking back and forth from pasture, to barn, to water, daily. In some circumstances, we will want to make our paths 'sacrifice zones'. We accept that these areas will suffer damage to the pasture sod, but by focusing the impact in one small area, we maintain the health of our pasture sod elsewhere. In other situations, we can regularly rotate the location of the path, distributing the animal impact so that no area suffers permanent damage. Utilizing these two strategies for pathways enables us to maximize the health of our pastures overall.

Optimizing Soil Biology

One final consideration with soil nutrient levels is the critical role of soil biology in the availability of these mineral nutrients. Healthy soil bacteria and fungal populations make more minerals available to our pasture plants. Given the inexpensiveness of various soil inoculants, improving our pasture nutrient availability with microbiology is always a good investment. This is an area where Biodynamic farming has a lot to offer the modern dairy farmer.

Ehrenfried Pfeifer's book *Soil Fertility* is an excellent resource for understanding the ways that healthy soil biology can dramatically increase mineral nutrient levels in our forages. Biodynamic sprays, such as Pfeifer Field Spray, will inoculate our soils with a rich collection of bacteria and fungi, optimally tailored to improve soil function. Pfeifer Field Spray will cost a fraction of phosphate fertilizer, and in many cases will have a comparable benefit. The critical variable for the farmer is the amount of phosphorous being utilized by our pasture plants, not the total amount of phosphorous in our soil itself. Pfeifer Field Spray works by increasing phosphorous availability, rather than total soil phosphorous. Pfeifer Field Spray is available commercially through the Josephine Porter Institute in Virginia.

Soil biology can also be improved through applications of compost extract. A relatively small amount of perfectly finished compost is soaked in water, and then diluted and sprayed on our pastures. The result is a microbial inoculation of good bacteria and fungal species that help to increase mineral availability to our plants. Compost extract is an excellent way to get the most bang-for-our-buck from valuable farm compost. In a pasture setting, where legumes and deep rooted pasture species are providing our nitrogen and humus needs, the most significant value of compost is in its microbial content. Compost extract allows us to efficiently spread these microbes throughout our pastures. Regular monthly spraying of compost extract is an economical means of improving our soil health and pasture fertility.

The way that we produce our compost will have a significant impact on the quality of its microbial content. In my experience, following the Biodynamic compost making technique will produce superior results. The specific technique of Biodynamic composting is discussed in detail in Pfeifer's book. The critical considerations are the use of animal manures, carbonaceous material, and clay soil, all inoculated with the specific Biodynamic Preparations. The Biodynamic Compost Starter preparations can be purchased through the Josephine Porter Institute in Virginia. The Biodynamic Preparations are individual compost inoculants, six of them, created by precisely composting certain microbially beneficial plants. The plants utilized are yarrow, chamomile, stinging nettle, oak bark, dandelion, and valerian. These six individual compost 'preps' are placed in the compost pile; they provide the optimal microbial influences to produce the best quality biological compost.

A more sophisticated tool related to compost extract is compost tea. Broadly speaking, a compost tea is created by taking the compost extract, and feeding the microbes in the extract with sugars and oxygen. As such, the microbial content is multiplied many times, greatly increasing the value of the compost tea for soil inoculation. The production of compost tea is a technical process, requiring specific equipment and techniques. The research by Dr. Elaine Ingham is highly recommend for anyone interested in maximizing the value of their compost for increasing farm fertility.

As dairy farmers, one final tool that we always have on hand is raw milk itself. Diluted ten-to-one with fresh water, raw milk is a tremendous microbial inoculant for our pastures. Regular monthly application of dilute raw milk will boost the healthy microbes in our soil. Raw milk also provides a nice foliar fertilizing effect on our pasture forages. Immediate positive growth responses from our pasture plants will be seen with raw milk sprays. Given that raw milk is a free and abundant resource on the dairy farm, its use in improving our pastures is an obvious part of economical, holistic, pasture management.

Soil Mineral Balance

Managing the soil nutrient levels in our dairy pastures is a difficult balance. Given the high cost of fertilizer inputs, it can be difficult to financially justify the costs of remineralizing our pastures. On the other hand, small mineral deficiencies can greatly compromise pasture yield, resulting in higher hay bills. Finding a balance between soil fertility and pasture productivity is a challenge for all pasture famers.

Holistic grazing management is an effective tool for building our soil humus, and raising our available nitrogen levels. Utilizing a combination of tap-rooted plants, fibrous rooted plants, and nitrogen fixing legumes, coupled with good grazing management, will result in steady improvements in these aspects of soil fertility. Application of organic nitrogen fertilizer is an unnecessary and unthrifty use of farm resources. Management intensive grazing alone should be adequate for restoration and maintenance of optimal nitrogen and humus levels in our soil.

Phosphorous, and to a lesser degree calcium and sulfur, are more challenging fertility needs. The cost of phosphorous fertilizers, either organic sources like soft rock phosphate, or refined sources like mono-ammonium phosphate, can be difficult to justify on a small dairy farm. Although focused on gardens and not pastures, Steve Solomon's book *The Intelligent Gardiner* contains an excellent discussion of the methods and value of phosphorous fertilization in healthy soils. If our phosphorous levels are severely deficient, fertility problems for our herd will likely emerge. In this case, the cost of phosphate fertilizer is certainly more economical than the losses incurred from poor calving rates.

Phosphorous is a critical nutrient needed in fairly large quantities by dairy cows. An average dairy cow will uptake 35 pounds of actual phosphorous annually. Deficiency in phosphorous will result in fertility problems, compromising our seasonal calving schedule, and resulting in loss of income from infertile cows. In a short-term emergency, phosphorous can be provided directly to our cows as a mineral lick, although it is much better to provide adequate phosphorous through the plants in our pastures.

Phosphorous is a stable mineral in the soil, so that the benefits of phosphorous amendment will last for many years. Phosphorous fertilization can be seen as a long-term investment in the fertility of our farms. In New Zealand, intrepid dairy farmers apply phosphorous to their pastures at the conclusion of especially profitable years, preferring to incest in their soil fertility rather than paying a higher tax bill for the year.

Calcium and sulfur levels in our soil will greatly affect the productivity of our pastures. Soils deficient in these two key elements will struggle to grow healthy legume plants, resulting in less nitrogen fixation. Low nitrogen levels will result in poor pasture yields. It has been said by wise farmers for over a century that it is much cheaper to purchase calcium lime or gypsum, than to buy more land or hay. This is even truer today, than in the past. The good thing about calcium and sulfur amendments is that they are comparatively inexpensive. The bad thing is that large quantities, especially with calcium, must be applied. Fortunately, calcium and sulfur are long-term amendments, which will benefit your soil for decades.

Micronutrients are more important than their name suggests. Deficient levels of copper, zinc, and boron especially, will result in unthrifty and unhealthy livestock. These nutrients are only required in small amounts, but they are critical for reproductive health and strong immune systems. When it seems cost prohibitive to apply these minerals to our soils directly, a good solution is foliar feeding of our pasture forages. Because these are micronutrients, we can effectively apply them in dilute foliar sprays, boosting the nutrient levels of our forages so that our cows receive adequate amounts in their diet.

Over-Seeding Permanent Pastures

In our permanent pastures, periodic seeding is another tool for progressively diversifying the species balance. Generally, we will not need to worry about sowing seeds of plants that already have good populations in our pastures. Our rotational grazing management will generally ensure that a few individual plants go ungrazed, and manage to develop seed heads in the pasture. This will provide a large enough residual seed population to maintain healthy plant populations throughout our entire pasture.

Seeding our pastures is most useful with short-lived perennials that need to be constantly renewing their populations. This is particularly true with red clover. Red clover will grow well for several years, and then naturally die out. Periodic application of small amounts of red clover seed will ensure a healthy stand of forage year after year. Seeding rates can be quite low, nothing like the recommended amounts for establishing a new pasture. The key is regular sowing each year to maintain a strong population. New perennial species that we wish to establish, such as chicory or plantain, should only need to be sown once in their initial establishment stage. Once they are growing in our pastures, these perennials should live long and freely reseed. We can gather their seed heads to disperse to other areas of our pasture as needed over the years, to broaden their populations on our farm.

A hard lesson learned over the years is that simply broadcasting seed into the pasture is a much better way to feed the birds, than it is an effective technique for establishing new plants. Many, many times, just as soon as I spend valuable farm capital on pasture seed, and broadcast it across my pastures, vast flocks of birds descend on my fields. I watch in dismay as migrating birds mercilessly consume every last seed I spread. Fortunately, after many failures, there is a better way.

The best way to establish new plants is to sow seeds in concert with our cows grazing rotation. Each day, when our cows are rotated onto fresh pasture, spread seeds on just that area. As the cows are grazing that fresh pasture, their hooves will be pressing the pasture seeds into the soil. With each new move of our cows across the pasture, fresh seeds are broadcast and trampled into the earth. Sowing small areas at a time does not attract birds, so little seed is wasted. Timing this operation to coincide with good rainfall will greatly increase its success. The elements needed for seedling germination in an established pasture are viable seeds, pressed firmly into the soil, and provided with moisture. When we provide all three requirements, we are nearly guaranteed success.

A second way to help establish new plants in our pasture is to sow seeds just before the first heavy snows of winter. This is not always possible, due to variations in climate. But if we can broadcast our seeds, and have them immediately be covered with deep snow, birds will not be able to eat the seeds. The freezing and thawing action of the soil over winter will gradually work the seeds into the soil surface. Come spring, the seeds will be held in the earth, and provided with ample springtime soil moisture. This technique is a bit more fickle than allowing our cows to trample the seeds into the soil, but it is still a useful method of over seeding established pasture.

Purchasing pasture seed is an area where thorough research will uncover better and more economical sources of seed. Locally adapted varieties are always good. Finding reputable seed dealers with good wholesale pricing will save a lot of money. Especially with less common species, such as chicory and broadleaf plantain, there may be huge variations in pricing between different seed companies. Careful research will uncover economical sources for all your pasture seed needs.

Irrigation

The use of supplemental irrigation in our pastures to ensure dairy quality pasture is a regionally specific situation. In areas where pasture irrigation is commonly utilized, there are two basic techniques, flood irrigation and overhead sprinklers. Both methods have their own pros and cons, and local preference for one or the other will dominate the decision-making process. In general, sloping lands with good water retention are more economically irrigated with flood irrigation. Flat land or sandy soils are more conducive to overhead sprinklers.

When irrigating our pastures, it is imperative that cows are not allowed access to a given area as it is being irrigated. The pasture sod becomes completely saturated during irrigation, and will be severely damaged by hooves when wet. As such, we will need to manage the timing of our rotational grazing in concert with our irrigation cycles.

Ideally, we will irrigate a pasture immediately after it has been grazed. Irrigation water will provide a rapid boost in plant growth rate, helping our pastures to regrow quickly after grazing. The pattern of moving across the land, grazing and then irrigating, creates a beautiful rhythm in management intensive grazing.

In many locations, irrigation water availability declines rapidly in the fall. Many pastures will see their final irrigations in August, or even earlier. Once our irrigated pastures lose water for the season, their growth rate will come to a near standstill. Of course, soaking rains will more than compensate for the loss of irrigation water. In fact, it is often said that an inch of natural rainfall is worth the equivalent of three inches of irrigation water. Nevertheless, understanding the relationship between pasture irrigation and plant growth will enable the dairy farmer to maximize the productivity of their pastures. Supplemental irrigation is just another tool that pasture farmers can use to better achieve their performance goals.

10. HAY, WINTER FEEDING, AND COW MINERALS

A Farm without a Tractor

Owning a tractor is one of the most economically destructive management decisions available to a small farmer. With a system of cattle feeding that is based on cows directly grazing the pasture, a tractor is a financial burden that cannot be justified. The holistically managed dairy herd, with its emphasis on minimizing hay consumption, can and should abandon any notion of growing its own hay on the farm. Haymaking equipment is tremendously expensive, to purchase, to operate, and to maintain.

Growing one's own hay rarely makes economic sense. Raising your own hay and relying on local contract farmers to cut it for you is a sure path to disappointment. Rarely will other farmers be ready to cut your hay when it is ready to be cut. Quality compromises will be the norm. Considering the feed needs of thousand pound bovines, cutting hay by hand is about as practical as counting grains of sand. Given these realities, it is assumed that all small dairy farmers will be looking to purchase their annual hay needs. Viewed holistically, buying hay is by far the most economical and practical solution.

Hay that we purchase will likely come from a large farm, which is completely dependent on mechanization. There, the financial burden of haymaking machinery will be justified by growing and cutting hundreds or thousands of acres of hay on an annual basis. Even in an era of 'expensive' hay, it will always be cheaper to buy hay than to own and operate haymaking equipment.

With no hay to harvest, no grain to plant, and no fields to plow, a tractor has minimal value to the holistic dairy farmer. For the annual task of clearing out manure and woodchips, hiring a local operator for an afternoon will be money well spent. Owning a tractor, with the antiquated attitude that it will 'find' all sorts of useful jobs on the farm, is a solution desperately in search of a problem.

Other mechanized equipment, like mowers and ATVs, are similarly uneconomical on the holistically managed small dairy farm. Walking through our fields, the farmer notices many details that are lost when zooming past on an ATV. Again, without vast acreage to cover, there is no need to mechanize movement on the small farm. Cows themselves are the best mowers, converting grass into milk and meat, without fuel costs or regular mechanical repairs. In this modern era of machines and gadgets, resist the temptation and social pressure to mechanize the farm. Your annual budget, and its more robust profits, will be all the proof necessary.

Hay Feeding, Handling, and Economizing

The ideal of a zero-hay dairy operation should be the aspiration of every small dairy farmer. In reality, almost all dairy farmers will never achieve this lofty and ambitious objective. As discussed above, pasture management techniques to minimize the window of hay feeding are critical to the economic sustainability of you dairy farm. Hay is expensive. It is expensive to buy, expensive to store, and expensive to feed to your cows. Careful consideration of your hay feeding needs and methods is critical for every dairy farm.

When the dreaded day arrives that your pastures have been completely grazed out, hopefully sometime in early winter, it is time to begin feeding hay. This moment should have been planned for months in advance. Hay is best purchased in the summer, fresh from another farmer's field.

Hay is a general term that describes any dried plant material. In actuality, there is a substantial difference between different types of hay. Alfalfa and grass are not nutritional equals, not even close. Alfalfa hay is vastly more nutrient and mineral dense than grass hay. Strangely, prices on the hay market frequently do not reflect this discrepancy in actual feed value. As such, the thrifty dairy farmer, knowing that alfalfa hay has a much higher feed value than grass hay, should always seek out alfalfa hay as their preferred winter feed.

Alfalfa: The Best Fodder

Alfalfa hay can vary greatly in nutritional composition depending upon the circumstances of its harvest. Conventional farmers want green alfalfa, harvested before the plant has begun to bloom. This is the stage at which the alfalfa has maximum protein levels. Green cut, 'dairy quality', alfalfa hay contains so much protein that the cow cannot assimilate it all. The cow has to expend energy to excrete the excess protein through her urine, wasting this feed value and negatively affecting her health. Conventional farmers 'solve' this problem by feeding 'dairy quality' alfalfa hay mixed with high carbohydrate grain, to create some sort of cow rocket fuel. For reasons already discussed, this is not our objective in cow husbandry. We are interested in feeding the cow a more natural diet that is more conducive to her health and longevity.

The best alfalfa hay for our purposes, is cut in full bloom, when the minerals in the plant were at their maximum, and the protein concentration has declined to more balanced levels. Fortunately for us, conventional farmers typically do not want alfalfa hay in full bloom, so it is generally easy to purchase. During the busy summer haying season, large hay farmers inevitably end up with a good field of alfalfa that did not get cut on time. This field of full-bloom alfalfa is perfect for us. The hay farmer is delighted to find an eager buyer for his 'past-prime' hay. When we come asking for alfalfa hay, cut in full bloom, we frequently find an eager seller. This reality makes our task of sourcing alfalfa hay at an economical price much easier.

The quality of the hay you are purchasing is more important than any other factor. Hay must appear bright in color, and be free from any mold or decay. Producing quality hay is not an automatic process. Cutting, windrowing, and baling at the proper times are critical processes for ensuring a quality product. Before purchasing, pull out a small sample from a bale, and check to see that the hay smells fresh, and is not dusty or moldy. Make sure that in addition to the plant stems, that there is plenty of leaf, as the leaves are where the nutrition is. Never accept any hay that has powdery mold residues. Your hay should be high enough in quality that you would not hesitate to make your wife a cup of tea with it.

If straight alfalfa hay is not available in your region, your next best bet is some type of legume/grass mix. Be sure to ask lots of questions about what you are purchasing. In beef country, 'cow hay' refers to basically anything dry that is in a bale. This is not good enough for your dairy cows. Similarly 'horse hay' is typically anything that is mold-free, but generally of low feed value, coming from over mature grasses. You want 'excellent quality hay', no exceptions. Hay salesmen are, well, salesmen, and you will need to inspect the hay and evaluate for yourself in most circumstances.

In a worst-case scenario, with no alfalfa hay or mixed legume hay available, early cut, first cutting grass hay will make a passable alternative. Grass hay must be cut before the summer solstice for it to have any chance of retaining good nutritional value. The grass hay must be primarily leaves, with an absolute minimum of stem material. Know that grass hay is much lower in nutrition and minerals, and that you will need to feed more of it to get the same value for your animals.

Choices with Bale Types

Hay is typically baled in small square bales, large square bales, or big round bales. Each has their own pros and cons. Small square bales are often the most easily available, especially in small quantities. The biggest drawback to small bales is the hassle of handling them in feeding. They are only small in a relative context. Once you spend a winter hauling around 60 plus pound bales of hay, you won't think of them as small anymore.

Large square bales are the most efficient way of handling and feeding hay. Particularly if you need any quantity of hay whatsoever, large bales are much more practical. A large bale can hold anywhere from 1000 pound to a full ton, per bale. When you are feeding out 40-50 pounds per day, per head, you will quickly see the merit in large bales. Ideally, large bales will be stacked under your hayshed, the strings cut one bale at a time, and the hay forked by hand, directly into an adjacent cattle feeder.

Purchasing large bales in bulk will generally get you a much better price on your hay. Stored properly, hay will last for many years without any loss of feed value. It follows that purchasing a full semi-load of hay, so long as you have adequate storage capacity, will be the most economical option. Most semis trailers can transport twenty-eight, one ton bales. This will likely be enough hay to last multiple winters. Purchasing in bulk will get you a wholesale price from the hay farmer, and will save on the hauling costs. Even if you need to go outside your immediate locale to purchase hay wholesale, you will likely come out ahead purchasing by the ton.

Big round bales are distinctly different from the square bales. They are generally fed whole to the animals, with a feeder setup around the outside of the bale. The downside to this is that you will need machinery to move each round bale into position for feeding. Round bales cannot be easily fed out with a pitchfork, as the hay is wrapped in a spiral and does not flake off nicely like a square bale. One significant advantage to the round bale is that in arid climates, they can be stored without cover. Round bales shed moisture relatively well, and can be stored stacked on their side, exposed to the elements. If you have a tractor but no hayshed, this may make round bales an attractive option.

Round bales work well for feeding directly in the field. Since you will need to use machinery to pickup each bale and bring it to its feeding location, you can choose to designate your feeding location out in the pasture. The bits of wasted hay and concentrated cattle manure can really help to enrich poor spots in your pasture during the winter hay feeding. Of course, depending on the initial quality of your pasture, you may risk doing more harm than good from the hoof compaction and tractor traffic though a potentially wet winter pasture. But for working to improve poor spots in an established pasture, direct hay feeding in the pasture can be an excellent technique.

Locating a central hayshed between your cow and calf yards is generally the most efficient setup for hay feeding. One large roof can shelter your entire hay crop. Cows can have a feeder on one side of the structure, and calves on the opposite side. Laying down a thick layer of woodchips under the feeders will help to minimize mud and maximize retention of manure nutrients. As the animals consume more of the hay, the feeders can slowly be moved inward, minimizing the distance you need to toss hay to the feeders. At the end of the hay-feeding season, the manure-laden woodchips can be scooped up and composted. This barnyard compost will be an excellent amendment for your garden or orchard, and will also be the source of valuable compost teas that you use to inoculate your pastures with healthy soil microbes.

When Pastures Fall Short

There may be times, especially in the early years of pasture development, that pasture alone will not support optimal cow health and nutrition. In this situation, it is proper animal husbandry to supplement our cows with alfalfa hay. Good stewards of the herd cannot allow their animals to suffer inadequate nutrition. Alfalfa hay, with its high levels of energy, protein, and minerals, is a perfect complement to sub-dairy quality pasture.

Feeding ten pounds of alfalfa hay, per cow, per day, in addition to her regular pasture rotation, will have a dramatic effect on animal health and performance. Ten pounds of alfalfa hay is certainly not an adequate ration of its own, it is merely a supplement to ample quantity, but low quality pasture. The alfalfa hay will balance out the lack of energy in mediocre quality pasture grasses, creating a decent total ration. Total intake of dry matter will still need to be in excess of 40 pounds daily, anything less than this will not enable adequate rumen fill, and the dairy cows will be suffering from acute malnutrition.

The best possible pasture supplement is alfalfa hay, sprayed with molasses. The alfalfa hay provides the protein, and the molasses provides a significant boost of energy in the form of sugars. Both alfalfa and molasses are high in minerals. Feed grade molasses can be purchased in five gallon buckets, diluted with water to an appropriate consistency, and sprayed over the hay at the time of feeding from a simple plastic spray bottle. This results in a truly exceptional feed supplement for times of reduced pasture quality.

Hopefully, with a good grazing rotation, pasture quality will improve over time, making such daily hay supplementation unnecessary. Nevertheless, the dairy farmer has to work with what he has, and a little bit of good alfalfa hay can go a long way towards good stewardship of the dairy herd.

The Hay Yard

A hay yard is a large area, covering a minimum of 300 square feet per cow. Its primary function is a place for feeding hay during the winter. It should be located between the grazing pastures and the milking parlor. The ideal hay yard incorporates a good windbreak along the windward side, a group of shade trees, and a small section of covered shelter, accessible during the worst storms. It will also require a reliable, year-round source of drinking water for the cows.

Shaping the hay yard in a rectangle, rather than a square, will make the hay yard an efficient location for sorting animals within the herd. When calves are nursing half of the day, it is necessary to bring the entire herd into the hay yard, and then to separate the cows from their calves. Locating your dairy barn between the hay yard and calf yard will allow individual animals to be easily isolated into the individual cow stalls, and then calves moved on to the calf yard.

In time, cows and calves become trained to the system, and individual attention will be unnecessary. It often looks just like moms dropping their kids off for daycare. Yes, there can be some separation anxiety, but moms look forward to a break from their offspring, and the young ones enjoy time on their own with their peers. Good design reduces stress by minimizing confusion for our animals. Once a routine is established, gates are simply opened and closed, and cows and calves quickly learn who goes where.

Sturdy fencing between the hay yard and the calf yard is an absolutely necessity, otherwise the strong mothering instincts of cows will cause great damage to the fencing. Once a cow learns that she is strong enough to smash a light wire fence, she is an empowered bovine, and may cause quite a bit of trouble for years to come. A double fence, setup so that cows and calves do not have direct access to both sides of a given fence, is highly recommended.

Use sturdy tubular steel corral panels along the cow side of the fence, leaving a 10-20 foot lane of neutral space, and then a fence constructed of welded wire cattle panels along the calf side. These fencing materials are desirable because they are easily reconfigured, and can be supported with easily movable metal T-posts. Fixed fences of stretched wire never seem to be located in the right place, and their permanent location makes other farm tasks such as mucking out manure more challenging.

The modular components of T-Posts, tubular steel corral panels, and welded wire cow panels will facilitate strength and mobility in our cow yard layout.

The Calf Yard

The calf yard is essentially a hay yard in miniature. Calves will do fine with less area than cows, maybe half the space per animal on average, or roughly 200 square feet per calf. The calf yard will need substantial shelter, and protection from both wind and snow should be excellent.

Locating the calf yard closer to the family home than the hay yard is desirable, as such an orientation will deter predators. Positioning the calf yard toward the home is also enjoyable for the increased human interaction with the calves. The more time and contact we have with our dairy calves, the more sure their temperament will be as they mature.

In the calf yard, the calves need access to fresh water, and pasture for grazing. The calves will not eat much during this time; mostly they will lie around resting. But they do need to have pasture available, though it can be a relatively small area. The point is that the calves are not to be locked up in a barn or a bare dirt lot. Their yard should be at least partially vegetated with grass. In a worst-case scenario, a small amount of grass hay could be provided to the calves. However, remember that at this point in their lives, they have never seen their mothers eat hay, and will be disinclined to do so themselves. A small amount of grass in their yard will allow them to continue practicing their grazing, imitating the skills that they have observed their mothers doing out in the pasture.

Woodchips and Manure

The best material for covering the ground in the cow yard is wood chips. Wood chips minimize mud, in an area that will see heavy impact throughout the year. The wood chips remain clean and dry, so that herding cows and calves through the yard is not a slippery and stumbling affair.

Wood chips soak up the cows' urine and manure, absorbing valuable nutrients so they are not lost into the ground. During the hay-feeding season, a large amount of fertilizer is captured by the carbon-based woodchip material. Carbon bonds with nitrogen and other nutrients, enabling it to be composted, and ultimately recycled onto gardens or orchards.

In mid-summer, the previous year's woodchips are shoveled out, and replaced with a fresh surface layer. Assuming the small dairy does not own large machinery like a front-end loader, hiring a machine for a few hours makes the most sense. We can plan a day where the loader clears out the yard in the morning, the dump truck arrives with a fresh load of chips at noon, and the loader then spreads out the new wood chips that afternoon. Small skid-steer type machines are the most efficient for this task.

Importing large amounts of woodchips annually, and relying upon hired machinery, is the most efficient and cost effective management system for the small dairy. The large, woodchip-based compost piles that we build each year provide excellent foraging for hens, reducing farm costs for purchased hen feed. Given a year to fully break down the carbon in the wood chips, this compost is excellent for use on gardens and orchards. The application of Biodynamic compost starter will facilitate complete breakdown of the wood chip material, and maximize the fertility value of our barnyard compost.

Drinking Water for Cows

Cows are perfectly adapted to walk long distances to their water. There is no reason for dairy farmers to disrupt this natural reality. Going for a leisurely afternoon walk is likely as healthy for cows as it is for humans.

Providing water in the pasture creates great expense in piping and stock tanks. Watering locations in the pasture will quickly become degraded mud holes. Maintaining a permanent watering spot, conveniently located adjacent to our barn and cow yard will produce excellent results. Woodchips should be spread thickly around the watering trough location to minimize mud and encourage cleanliness.

The water supply in the cow yard is most challenging during the winter months. Having a moveable stock tank is recommended to be able to rotate the heavily impacted area immediately surrounding the watering area. Locating a frost-free spigot in the neutral zone between the cow yard and the calf yard is a good design. That way, bored cows cannot decide to smash the spigot. A rubber hose can be easily run into either yard as needed. This is much preferable to a permanent location for the water tank with a dedicated spigot in a fixed location.

Quality of water is much more important for our cows than convenience of access. It is important that our water source for our cows is clean and sanitary. A slowly overflowing stock tank, with piped discharge, is an excellent solution. An overflowing water source will not freeze easily in winter. Stagnant water is always inferior to running water, from a health standpoint. A good source of fresh running water, with enough drinking space for several cows, will encourage ample water consumption, which is critical for milk production.

When planning a watering spot, choice of water vessel is important. Rubber watering tubs are great because they will not be broken by the cows, and when water inevitably freezes solid, the ice can be broken out without breaking the tub. Little Giant is an excellent brand, and I have yet to destroy a single rubber watering tub of this design. In severely cold winter areas, where daytime temperatures stay below freezing for weeks at a time, an electricity source will be desirable to keep stock tanks from freezing solid.

Never underestimate the amount of water needed by dairy cows. Plan on 40 gallons per day minimum for a lactating milk cow, and 10 gallons per day, per head, for all dry stock. Hauling water, whether by bucket or by truck, will quickly become an onerous task.

Natural Shelter

Managing areas of forest or hedgerow to provide shelter for our animals is a valuable asset on a small farm. Large areas are not required to provide significant protection and shelter from driving winds and storms. Unless you live in an absolute rainforest, protection from rain is not needed for cows. But strong winds, in combination with rain or snow, are not healthy for our animals. Managing the farm vegetation to provide shelter from the elements will increase the amount of time that our animals can stay out at pasture. Time spent out at pasture is good for our animal's grazing, good for our pastures' fertility, and good for our economy of feeding purchased feeds.

The key concept to aim for with our vegetative shelterbelt is maximum edge area. We do not need or want large areas of trees, as this just takes away from our available pasture. Having many narrow strips of trees and shrubs provides numerous sheltered locations, while taking up a very small total acreage. Selectively cutting existing woodlands to shape more pasture and better vegetative shelters is an important strategy. Wooded areas fifteen to twenty feet wide, managed for maximum vegetative density, will be the most efficient shelters for our animals.

In the cow and calf yards, windbreaks are likely the most important consideration for shelter. Windbreaks can be provided most efficiently by utilizing the side of a building that is located to the windward side of the cow yard. Otherwise, board fencing is very effective at making a windbreak for the cows.

Remember that cows will try to rub up against most anything, so your windbreak needs to be sturdy enough to withstand the considerable force of a thousand pound cow. Thick stands of evergreen trees, like spruce and fir, can make excellent windbreaks, though their establishment would require significant time.

Winter shelter in the confines of the cow and calf yards is an important consideration. Unless living in an extreme climate, such as would not be suited to natural farming of dairy cattle anyways, direct protection from cold is not necessary. During the worst winter weather, some dry area should be provided. This is best done by constructing an oversized hay shed, with some portion of it accessible to the cows during severe weather. This area should be closed off during decent weather; otherwise the cows will make it their preferred home and soil it excessively. This sheltered area can be as small as 100 square feet per cow.

When the weather is really bad, opening a set of gates to allow access here will ensure the health and contentedness of your cows.

In especially hot and sunny climates, having savannah-like islands of trees in our pastures will improve animal comfort and performance. Islands of shade in a hot sunny pasture will give our animals comfort, and they will yield more milk as a result of their gratitude. Existing woodlands can be cleared in a careful way that leaves small clusters of trees, dotting the pasture. Three to five good trees tightly grouped together will provide cool shade through a majority of the day.

> *Establishing shelterbelts and savannah islands is challenging with the natural grazing habits of the cow. After observing cows for many years, I have concluded that cows need grasslands to survive, and have an evolutionarily based instinct to create more grassland. Cows seem to have an abject contempt towards young seedlings, and will trample, gore, and destroy them with relish. If you are trying to establish shelterbelts in your pastures, physical fencing barriers will be necessary until the trees are quite large. I have watched a cow grasp a seedling tree with her tongue, and rip it out of the ground like an elephant. I have observed a cow walk up to a young tree, lean against it, and sit down directly on top of it, snapping its trunk in half. This is to say nothing of the delight cows demonstrate when smashing tree branches with their horns. Cows love grass, and have many successful strategies for farming their own pastoral grazing paradise.*

It will take many years to establish a mature shelterbelt, and every single day during that establishment phase, your cows must be physically restrained from direct access to the young trees. A decade from now, your cows will love the mature trees for their shelter and shade. Additional species diversity will eventually thrive in the shade of your trees, providing many unique medicinal plants for your cows to utilize. But this reality will take many years to develop, and your careful stewardship throughout will be imperative.

Worming and Woodland Self-Medication

Cows have an amazing instinctual understanding of the many medicinal properties of wild plants. For this reason, maintaining a small wild woodland area, that contains a wide diversity of plant species, will enable the cow to self-medicate when needed. Periodic seasonal access to this woodland will cure many health problems before the farmer can even notice them. One of the many medicinal issues that cows seem to be very effective at addressing in this way is intestinal worming.

Worming is the periodic process of cleansing the digestive tract of any parasites. Traditional chemical wormers are highly toxic to the cow, destroying as many beneficial organisms in the digestive tract as parasitic ones. The regular systematic use of chemical wormers is completely unnecessary, and certainly harmful, for pasture-based dairy cows. Allowing our cows access to healthy pasture, coupled with periodic access to a good diversity of native medicinal plants in our hedgerows and woodlands, will work to promote optimal digestive balance for our cows without additional worming.

Salt and Minerals

Salt is an essential nutrient for all cattle. Providing a good quality salt source, complete with natural minerals, is a wise investment in the health of you herd. Conventional cattlemen use 'mineral salt', which is a chemical compound that includes salt and refined minerals, formulated for cows. This conventional mineral salt is the human equivalent of iodized salt and vitamin supplements. It is questionable how much of the mineral content is actually assimilated. As in human nutrition, it is wise to feed our cows something akin to sea salt, which is full of trace minerals in a more natural form.

Vast deposits of salt occur in the Western US, deposited by the ancient inland seas that once covered this region. A particularly rich salt deposit exists in Utah, home of Redmond Real Salt. This salt, produced both in human table grade and livestock grade, is rich with minerals, like sea salt. Real Salt can be conveniently purchased in salt lick blocks for our cows. It may cost marginally more than conventional mineral salt, but it is a superior product that is an economic way to ensure our cow's complete micronutrient mineralization. All animals in our cowherd should have access to a salt lick every single day of the year.

Additional mineral supplements should be approached with caution. On one hand, cows seem to have a tremendous aptitude for sensing their own bodily needs, and consuming only the minerals that they need. Experiments providing cows free-choice access to a variety of macro minerals, like phosphorous, magnesium, and calcium seem to demonstrate that cows understand how to formulate their own supplements far better than farmers could. In situations of acute mineral deficiency on the farm, this is probably a wise practice. In most circumstances, a program of targeted pasture fertilization, coupled with wintertime feeding of mineral rich alfalfa hay and real salt, will naturally produce excellent animal health.

Kelp and Herbal Supplements

Kelp is a remarkable source of micronutrients, presented in a biologically available plant form. Dairy veterinarians have found that no other single management decision is more improving of holistic herd health than feeding kelp. With kelp, beyond the easily explained micronutrient benefits, there seems to be something more than the sum of the parts. Allowing your cows regular access to free-choice kelp is the best health insurance you can buy your herd.

Kelp consumption is directly related to the health of your animals. Particularly when you have newly purchased animals, who may not have been managed optimally, they may simply gorge themselves on dried kelp. Fortunately for your pocketbook, this consumption will taper off, and well-mineralized cows will not eat an excessive amount of kelp. Dried kelp is best purchased in 50-pound sacks, and should be seen as a basic staple of cattle nutrition and preventative care. It can be mixed 50-50 with granulated Real Salt, and fed in large rubber tubs or buckets.

Dried herbs have the same tonic and medicinal effects for cows as they do for humans. The biggest challenge with utilizing dried herbs for our cows is the cost of providing herbs in the quantities that our cows appreciate. Dried stinging nettle, and dried raspberry leaf are two herbs that merit collection for your cows. Fed late in pregnancy, these herbs promote healthy calves and easy deliveries. Fed again just before and during breeding season, they promote ready conception. Both nettle and raspberry can be easily harvested in large quantity, with minimal time invested. For more information about herbs and their role in cattle health, see *The Complete Herbal Handbook for Farm and Stable* by Juliette De Bairacli Levy.

11. BREEDING AND CALVING

It is always a bit amusing explaining to city folks, that cows only give milk after giving birth to a calf. Managing our cow's breeding is fundamental to controlling the flow of milk in our dairy.

Breeding our cows may primarily function to get them to produce milk, but gradual improvement of our herd's genetics is an important secondary benefit. Implementing a breeding program that improves the genetic quality of our cows is highly beneficial for the long-term prosperity of our dairy farm.

Heat and Conception Identification

The first step towards getting our cows bred is the identification and monitoring of their heat periods. The bull has tremendous powers of observation and identification regarding the fertility of his cows. The farmer, though a bit more blunt in the senses, needs to develop the ability to recognize heat cycles and signs of conception in his herd as well.

The fertility cycle begins when the cow comes into estrus, called 'heat'. As the cow comes into heat, several notable changes in behavior and demeanor may be observed. Of course, all individual cows react uniquely to their heat cycle, so these are just general guidelines that may or may not apple to each and every cow.

The biggest giveaway that a cow is in heat will be that other cows in the herd will mount and ride her. If the cow is not fully in heat, she will not stand still when mounted. When she is at the peak of her heat, she will be in what is called 'standing heat'. This is where the phrase, "I won't stand for it" comes from. A cow at the peak of heat will most definitely stand for it. She will also attempt to mount other cows at this time, but since they are not in heat themselves, they will not stand for being mounted.

Cows riding each other is a giveaway sign to any bull in the area that there is a job to be done. Riding is very obvious. Be forewarned, that cows will not limit their riding to each other. A very dangerous situation can arise when a cow spontaneously tries to ride the farmer. A thousand pound bovine crashing down on your back is not the indication of heat that you want.

Be careful around any cow that is coming into heat, she is at her most unpredictable and dangerous. Along those lines, cows in heat tend to act agitated. They may moo excessively, shake their heads aggressively, and you will wonder, "What got into that cow?" This is her fertility cycle manifesting itself.

Cows will often get a certain glint in their eye when in heat. They may stop and lock eyes with you. There is a certain sparkle of life that their eyes contain. This is the life force rising within them. Discharge from the vagina may also be present when the cow is in heat. More often, the discharge is most obvious after the cow has passed the peak of heat. It is as much an indicator that the cow just had her heat, than that she is still peaking in her fertility. Nevertheless, mucous or bloody discharge are definite indications of an active heat cycle.

The final indication of a cow in estrous is a drop in her milk production. When a cow comes into heat, her daily milk production will typically drop by a few pounds, for usually a three-day period. If she does not conceive during her heat, her milk production should rebound back to the pre-heat level. If she does conceive, her milk production will drop another few pounds, resulting in a drop of a half-gallon to as much as almost a gallon total decrease.

In a cow with a less robust constitution, her milk production will decline more severely. This is not a good sign; it indicates that her system is overwhelmed by the nutritional requirements of both lactation and conception. A small drop in milk production is normal, and will let us know that our cow has conceived. This redistribution of nutritional resources is a natural and healthy response to the increasing physical demands of pregnancy.

Timing the Pregnancy

Planning our cows' pregnancies begins with looking at a calendar, and identifying the ideal time in the spring for our calves to be born. We have discussed previously that we ideally want to have our calves born just before our average frost-free date, in order to ensure optimal pasture conditions for our pregnant and calving cows. We then count backwards, nine months, and mark that date on our summer calendar. This is the target date of conception for our cows.

As breeding season approaches in mid-summer, we will be documenting the cows' heat cycles. Meticulously marking down the dates of ovulation for each cow on a calendar will aid the precision of our management. Fertile cows will cycle roughly every three weeks, though again there is a range of healthy estrous that ranges from 18 to 23 days. During a given heat period, we typically have a 24-36 hour window of maximum receptivity, during which time we want to breed our cows, for the highest possible odds of conception.

Our cows are dynamic living beings, and their conception is not a mechanical event. We must remember that they will only come into heat every three weeks, so if we are planning an ideal conception date of July 25, but they ovulate on July 22, then the next possible date of conception will not occur until sometime around August 12. Because we have our records of their estrous cycles, we should be able to anticipate their fertility periods, and plan accordingly. As such, we may hedge our bets a bit, and breed that cow on July 22, because that will result in a better calving date than waiting until Aug 12. Good records facilitate good management.

There is no guarantee that our cows will conceive from their first breeding. We manage our herd for optimal health, nutrition, and fertility, so that our cows are highly receptive to conception. Nevertheless, there is still an element of chance in breeding our cows. When we are a little late breeding, as in our Aug 12 example above, and the cow fails to conceive, then we get pushed back again, and are looking at a conception date of Sept 3. Our plans for spring calving are now severely disrupted, from an initial planned date of April 25, to an actual calving date of June 3. This represents a significant loss of productivity for the dairy.

The above example illustrates the importance of good planning, and good fertility in our herd. We should accept that a given cow may not conceive on her first breeding, but cows that fail to conceive during their second or third heat periods are not acceptable. A seasonal dairy requires animals with excellent conception rates, and animals that do not satisfy this criterion over multiple seasons must be culled from the herd. Record keeping lets us track our cow's heat cycle and breeding performance, so that we can make productive management decisions about the composition of our herd moving forward.

Good record keeping concludes with marking the conception date on the calendar. Tracking the cow's gestation, the official due date is 285 days from conception. Since a full term pregnancy is anything plus or minus two weeks from that due date, we should mark our calendars for next spring, 270 days from the date of conception. It is much better to be prepared in advance, than to be surprised by an unexpectedly early calf. Additionally, given the massive size of a bovine calf, a slightly early birth is a real blessing for the cow's ease of calving. We both hope and plan for the earliest possible full-term pregnancy.

Choices for Breeding

Breeding the cows in the herd will involve one of three primary situations. The most desirable is owning a breeding bull in the herd. A second option is utilizing artificial insemination. The third choice would be borrowing a bull from a neighbor or fellow farmer. All three arrangements have their own pros and cons, which are now discussed in detail.

Breeding Bulls

The prospect of owning and caring for a dairy bull is a daunting thought for many farmers. Dairy bulls are powerful, dangerous, and potentially ill tempered. Managing a dairy bull is a serious undertaking, and should never be taken lightly. There are genuine risks to your fences, your buildings, and your personal safety. Bulls are good at causing a lot of trouble on the farm, as many an old farmer's tale will attest.

The one thing that bulls are the flat-out best at, however, is breeding cows. If you think your observations and records are meticulous, you have surely never compared notes with a breeding bull. They take their work seriously, and are consummate professionals. A bull uses his sense of smell to constantly inventory his cows' fertility cycles. He will know when their heat is coming, and when the moment of peak fertility has arrived. The bull will even be able to smell his cows a few days later, and content himself in the knowledge that the cow has conceived. Bulls are masters of fertility, and ensure the most successful breeding program possible.

The fertility power of bulls is the stuff of mythological legend. Religious cults all over the world have worshipped the tremendous power of creation that the bull embodies. A single bull ejaculation contains four to five billion sperm. That is some power of redundancy, since only one calf will be conceived. Between the diligence of the bull in following the cows' fertility cycles, and the potency of their ejaculate, bulls are profoundly effective at keeping their herds pregnant and prospering.

When you are in the same area as the bull, you must never turn your back on him. Your safety is always the first priority. Dairy bulls are inherently dangerous, and in traditional times when bulls were common on small farms, many good men were tragically gored through carelessness. Carry a sharp metal poker with you; an electric fence pigtail post works well. Waving your poker should immediately cause your bull to retreat, but never assume it will. If it comes down to it, and you find yourself under physical threat, do not hesitate to use great violence to protect yourself. Swinging your poker at the bull will cause minimal pain to such a large and thick hided animal. If the situation is truly dangerous, do not hesitate to stab your bull with the sharp end of the poker. Obviously, this is not your aspiration, but it may be necessary to protect your life. Gentleness is simply not a persuasive technique when dealing with an agitated and potentially aggressive bull.

As was discussed in the chapter on Holistic Herd Management, keeping the bull away from fertile females that we do *not* want bred is a serious consideration and challenge. Good fences are an absolute must. A combination of a physical metal wire fence, coupled with a strand of hot electric fence twine, is the most secure situation. A bull can easily jump a 54" barbed wire fence to get to a fertile cow. Placing a strand of hot electric fence wire on the inside of the wire fence will assuredly convince your bull that this fence has superpowers, and should not be messed with. It works. If a bull ever decides to test a fence, the battle is already lost. Good fences are an absolute necessity to contain the libido of a mature breeding bull.

Artificial Insemination

Artificial insemination is lauded by most conventional dairy farmers as the modern way to breed their cows, yet it has its drawbacks. Yes, artificial insemination (AI) enables the farmer to not have to deal with a bull. This is its most significant benefit. However, AI offers a pale shadow of the effectiveness in conception of a breeding bull.

The importance of prompt breeding in a seasonal dairy is fundamental. We need our cows to punctually calve in the spring, to maximize our utilization of pasture resources for seasonal milk production. Because there is no bull to identify the peak moment of fertility, it is very challenging to know when to best breed a given cow. This point cannot be overstated.

When we are in our desired breeding window, we will need to be absolutely vigilant in monitoring each individual cow's estrous cycle. Once we identify that the cow is in standing heat, she needs to be bred within hours. Artificial insemination equipment is expensive, and generally will not be owned by a small dairy farmer. An AI veterinarian will need to come out to our farm promptly, or we will miss the opportunity for our cow to conceive.

Commercial AI programs typically are conducted in concert with hormonal manipulation of the cow's fertility cycle. Veterinarian visits are scheduled weeks in advance, hormones are injected to control ovulation, and semen is inserted a specified number of days later. When we have only one cow to breed at a time, with a critical timeframe, we are not likely to be a priority for an AI veterinarian.

In contrast to the four billion sperm in a bull ejaculation, an AI straw contains only one-one thousandth the sperm population. As such AI does not have near the breeding success rate. AI programs typically measure their success by conception rates within a 90-day period. This is not nearly a tight enough window for a seasonal dairy.

Artificial insemination does allow for utilization of superior bulls, enabling farmers to use bulls from improved genetics to the ones found in their own herds. This can be a significant advantage for AI. If you have a poor quality genetic base for your herd, selecting bulls from within will have limited genetic capacity for herd improvement. The one caveat is that using AI limits our choice of bull type to those that are commercially available.

Seeking out bull semen that comes from dairies with similar objectives to ours is essential when relying upon AI to improve our herd.

It is also inherently problematic, as AI sires almost exclusively come from industrial dairy genetics. Locating AT semen from holistically managed pasture dairies will require extra work and research. Artificial insemination also compromises our self-sufficiency by putting us in a position of dependency for the reproduction of our herd.

Maintaining a Closed Herd

Selling animals is a natural consequence of breeding and selecting your optimal cowherd. Bringing new animals into your already established herd is a less desirable situation. Once you have a healthy, proven group of dairy cows, adding new animals from other farms introduces many potential health risks for your cows, and your milk. After the initial establishment period of your herd, you should strive to only build your herd from within.

When your herd is first established, you will need to test every individual for tuberculosis and brucellosis, plus any other cattle diseases that may be present in your region, or the region from which your new cows originate. These tests should be performed ahead of time on the farm you are buying your cows from, so that you have clear laboratory results giving a clean bill of health before the sale proceeds. Any positive results from disease tests are an immediate red flag, and you should look elsewhere for your cows.

Once you have a healthy cowherd established on your farm, you will not want to jeopardize the health of every individual in your herd by bringing in outside animals. It is not possible to test an incoming cow for every possible illness. One imported cow or bull could introduce a sexually transmitted disease, or could spread unhealthy intestinal pathogens throughout your herd. Consultation with your state veterinarian will alert you to the various cattle diseases that you should test for. Your state veterinarian will be especially concerned about diseases that may be introduced by animals originating outside the state. The state vet should be able to give you a clear list of tests to be performed, and the appropriate means for obtaining such tests. This is one instance where the council of a conventional veterinarian will prove very useful. In any but the most necessary of circumstances, the risks of importing new animals into your existing herd are not worth the possible rewards.

If despite the above cautions, new animals are to be added to the herd, and they have been thoroughly examined beforehand with a perfect bill of health, you will still want to quarantine them for several weeks upon arrival to your farm for observation. Keep the new animals physically separate from any animals in your herd. This will present management challenges, as you will need an isolated pasture in which to observe and monitor the new animals. Nevertheless, this is an important precaution to take when bringing in new animals to your farm.

At times on the small dairy farm, neighbors may request to use your bull for breeding their cow, or may ask you to board their animal for a short period of time. You may also consider using a neighbor's bull for breeding your cows. Neighborliness aside, it must be restated that the risks of bringing in outside animals to a raw milk dairy herd are not worth it. Politely explain the meticulous protocol that you follow to ensure perfect quality raw milk to your customers. The potential, even very small, of introducing a health problem into your herd, are simply not worth the risk to your cows or your customers.

Line Breeding

Cattle farmers practice a form of planned, selective inbreeding called line breeding to improve the quality of their herds. Line breeding works to genetically concentrate the desirable traits we have within our herd, so that they are more reliably passed on to successive generations.

The one rule that we must always follow in line breeding is that fathers never breed daughters. Centuries of cattle breeding have demonstrated that genetic degradation occurs when we allow this breeding pairing to occur. Otherwise, human concepts of incest and inbreeding simply do not apply to cattle. Brothers and sisters, mothers and sons, daughters and uncles, are all perfectly acceptable breeding combinations in a line breeding system.

When working with a given bull, it will be many years until we need to concern ourselves with unhealthy inbreeding in our herd. At a minimum, any given bull will be in service for three years. This is how long it will take his first female offspring to reach breeding age herself. Of course, if we are particularly committed to this bull, due to his outstanding quality, provision could be made to breed his daughters by other means, while that bull continues to service the rest of the dairy herd.

To begin the breeding program, acquire the best bull possible, from a dairy farm with comparable management style and production goals to our own. After this first bull, which comes from outside the herd, utilizing bulls from within our herd will produce superior results in genetic improvement. The powerful influence of local adaptation guarantees that bulls that are brought in from afar will never match the breeding bulls that are developed on farm. This is a critical point to recognize, as there is often a notion that outside blood would strengthen the genetics of our herd. Unless our animals are of particularly poor quality, this is simply not true. Working with the genetics present in our herd, passed through the lens of selection based on local adaptation, will develop our cowherd to its maximum potential.

The program of line breeding begins with our purchased bull, and progresses through successive generations. The purchased bull will, in time, sire our first homegrown bull. The homegrown bull will in turn sire our king bull. And finally, our king bull will sire our supreme bull, who will profoundly concentrate our best genetic traits to improve our herd. The same process that is used to develop our supreme bull will be repeated indefinitely to continue improving our herd. Following the principles of line breeding, we can now envision how breeding a closed dairy herd will progress over the generations.

For the first few years, we will breed our cows with our original purchased bull, and meticulously compare every bull calf to our original sire. It may take several years, but we will eventually produce a male offspring that is superior to his sire.

This outstanding bull calf should be born from one of our best cows; the kind of cow that we wish our whole herd was more like. This new bull will become the herd bull, and we will call him the homegrown bull. Unlike our original bull, which came from outside our herd, the new bull is fully half the genetic offspring of one of our best cows. This begins the process of concentrating our most desirable genetics. At this point, we will sell our original purchased bull, and move our breeding program forward with our new homegrown bull.

Over the next few years, we will carefully evaluate all offspring from our homegrown bull. Once again, we will come to select a superior bull calf, sired by our homegrown bull. We can call him our king bull, and he will replace our homegrown bull. We now need to select a queen cow for our king bull. Our queen cow will be a top quality female, sired by our original purchased bull. By now, she will be a tested and proven milk cow, with production records to demonstrate her productivity. To create our supreme bull, the king and queen are then bred together, and a superior bull calf is selected from this mating.

The supreme bull then replaces the king bull, and is used to breed our herd for the next three to five years. The supreme bull has a high percentage of the genetics of our best cows, and as such, will impart these traits strongly on his offspring. We do not leave genetics to chance, we stack the deck so that the traits of our best cows become genotypically dominant traits that are expressed by all future offspring. This is a textbook example of line breeding, where we can concentrate the genetics of our most desirable animals, over generations of breeding, and then use those concentrated genetics to improve our entire herd from within.

In time, our supreme bull will reach the end of his service life, as his daughters reach maturity and enter our dairy herd. Along the way, we will have been working to develop a replacement for our supreme bull. This next generation bull will come from one of our homestead bull's best cow offspring, now a proven producer herself, mated with our supreme bull.

The process of line breeding is even more potent if we can select a bull and a cow out of the same maternal lineage, to be mated together. Sometimes this is not practical, as genetic lineage is important, but it remains secondary in importance to the individual quality that the animals express.

By now, almost ten years into our line breeding program, our best cow will be a significantly better dairy cow that the original queen cow that we used to produce our first supreme bull. Even with a small herd of four cows and one initial bull, we can perpetually continue the process of line breeding. The process of line breeding thus continues, improving our herd from within.

Calving

In the last month before your cow is due to calve, she will get huge, full with her maturing pregnancy. During this time of maximum calf growth, the cow should have excellent quality lush spring pasture to graze. Alternatively, she could have top quality alfalfa hay to eat, although such dry feed lacks the laxative qualities that will help facilitate an easy calf delivery.

The first major indication that calving is near at hand, will be the calf descending out of the cow's belly and into her pelvis. In response to the internal pressure this creates, the cow will walk in a labored fashion. Slightly jarred, stilted motions will replace the smooth movement of her legs. The cow's vulva will become increasingly swollen and engorged. She will frequently reach back to lick her belly, a way of communicating with the calf in her uterus. She will be less social than usual, sometimes with a preoccupied look to her usually regal disposition. The cow is preparing for calving.

There is a certain temptation at this time to bring the cow into the barn, to observe her frequently. In general, this is not the best way to facilitate a smooth and easy birth for the pregnant cow. Assuming that we have timed our calving properly to the spring weather, there is no concern about the elements, for either cow or newborn calf. Thinking that the cow needs shelter for calving is anthropomorphizing at its well intentioned worst.

The best thing for the cow as she approaches labor, is access to the widest possible diversity of living plant species. Birthing is a time when cows make maximum use of their vast instinctive knowledge of medicinal plants. For example, given a chance, cows will almost always graze shepherd's purse while in labor. This is a member of the mustard family that cows otherwise ignore, but in labor, it serves some useful purpose for the cow. In different ecologies, no doubt other species fill this role for the cow. Allowing the laboring cow access to wooded areas with high species diversity gives them the natural shelter, privacy, and medicinal plants that work together to promote a healthy natural birth.

As the cow moves into active labor, the first major sign we see is a large amount of vaginal discharge. Generally mucous, and some blood, will be evident. The cow's hindquarters will be soiled with this discharge, a clear sign to the farmer that birth is imminent. As the cow's hormones take over, her typical consciousness is completely transformed. Most cows become completely unafraid of people in these moments. This is a good time to apply oil to the cow's vulva. Rubbing olive oil or any other moisturizing oil into the swollen vulva will help it to stretch without tearing during birth.

The next thing to watch for is the timid emergence of the calf's hooves. This is an important signal for the farmer, as it lets us know that the calf has presented with the proper orientation. Two tiny hoofs should be sticking out, neatly side by side, curling downward. This indicates that the calf is coming out front feet first, with its spine upwards. We can imagine a little bovine superman flying out of the womb; this is the correct presentation. By the time that the calf's feet have emerged up to the ankle joint, identification of this position of presentation should be obvious. Any position other than front feet first, with spine up, is cause for serious concern.

Until the dairy farmer has significant experience with calving and veterinary care, it is strongly recommended to call a professional cattleman or veterinarian if there is an unnatural presentation. The novice farmer can make matters far worse by rolling up his or her sleeves, and attempting to manipulate the position of the calf within the birth canal. The basic guideline is that if the calf is presenting properly, walk away and let the cow do her job. If the calf is improperly positioned, let it be, and call for professional help. You are more likely to cause harm than good attempting to intervene in a problematic delivery.

The moments in advanced labor, as the cow works to push out her calf, are truly awesome moments to observe. The feats of strength and endurance that the cow demonstrates are breathtaking. She will stand up and lie down, seemingly discontent, but in actuality performing a careful yogic sequence of movements designed to deliver the calf. In one moment her hind leg will be stretched outward, then the opposite front leg; the sequence of movements a profound display of inherited wisdom from the deeply maternal cow. It is always a profoundly beautiful and spiritual experience to watch a cow naturally give birth in a spring pasture.

The first time heifer cow deserves more thorough observation. Her vulva is much tighter, never having given birth before, so more frequent application of moisturizing oil is recommended. The older cows in the herd will have spent the preceding weeks coaching the heifer, silently explaining to her the secrets of birthing. During labor, a head matriarch in the herd may approach the heifer to encourage her. Let these interactions occur undisturbed. Cows are herd animals, and during these significant life milestones, they support one another in magical bovine ways. Cows are the best midwifes for one another.

Soon the shiny muzzle of the emerging calf will present itself. The nose should appear between the two front feet, emerging about the time that the front feet have exited just past the ankle joint. From this moment, the calf should be delivered within a half hour. The calf at this time is severely compressed within the birth canal. A smooth progression of labor ensures that the calf maintains adequate oxygen and blood flow in the moments before birth. Give the cow space, do not let anything disturb her, and allow her to complete her work.

If the calf has presented fully, with feet and nose emerged, but progress has stalled, the cow may be having a difficult time pushing out her calf. At this time, some assistance may be required, particularly in the first time heifer that does not have the same pelvic capacity of a mature cow. Additional factors that contribute to difficult deliveries are births occurring later in gestation, and bull calves in general. Both of these factors lead to larger calves, which are more difficult to deliver. When you find yourself in such a situation, the first thing to do is be patient. Place a call to your vet, alerting them that you may need assistance with a calving. Then relax, and let your cow work to resolve the situation.

There should be a steady ebb and flow of the calf's position, as the birthing contractions work to squeeze the calf out, and then relax. So long as movement is occurring, trust that the cow is making progress. If there were no ebb and flow occurring, this would be cause for concern. Get the vet to come out to the farm. The calf may need to be pulled. In time, with experience, this will be well within the scope of the skills of the dairy farmer. Learning directly from a professional is certainly recommended.

To help pull a calf, a rope will be cinched around the emerging hooves. Working in synchronization with the cow's contractions, force will be applied back and downward as the cow contracts. When her contraction subsides, release your pull. You will be pulling with all your strength; pulling a calf is an exhausting procedure. The largest part of the calf is their shoulders. Once you have passed this critical point, the hard work is done. Let the momma cow finish the delivery unassisted from here. When you help to deliver a stuck calf, a certain bond between you and that cow will always exist. It is a magical moment, feeling the calf finally pop free and slide out of the birth canal.

Nursing the Newborn

Immediately after birthing, the momma cow will often be so exhausted that she simply lies down for a few minutes. The calf too will be exhausted, lying motionless on the ground near its mother. Allow the cow and calf a few minutes to recover from the birth. Patience is always better than intervention. Soon, the cow should be up and licking the calf, pushing away the remnants of the birthing sack, and cleaning it of the amniotic fluid. If the cow, especially a first-time heifer, does not take an interest in her calf within about fifteen minutes, it is appropriate for the farmer to step in and encourage the mothering interaction.

Using a clean towel, vigorously rub down the newborn calf. This will clean the calf, help to dry its hair, and the massage will stimulate its circulation. Any interaction with the calf should trigger the momma cow's maternal instincts, and she will likely come over to protect her calf. Following your lead, the cow should begin licking the calf herself. With this stimulation, the calf will struggle to its feet. Falling and standing again, the calf will be exhausted by its first efforts to remain upright. Be patient, so long as cow and calf seem to be headed in the right direction, they will sort out their difficulties.

The momma cow licking her baby's rump stimulates the calf's instinct to nurse. It can be a hilariously awkward situation, the calf struggling to remain on its feet and find the teats to nurse. Keep your distance, and enjoy the comedy. There are several things we want to see, that let us know that ultimately the pair will find success. The cow should be licking the calf's rump, steering the calf into position near the udder. She will likely be alternately licking and mooing, anxious and agitated at the same time. The calf will be struggling to stay on its feet, nodding its head in an instinctive butting motion. The cow's udder may be so full that she is actually leaking milk. It will take a seeming eternity, but the calf will finally manage to latch on and take its first few sips of milk. Once this happens, the calf is set for a life of loving momma's milk.

Occasionally, you may need to squeeze some milk out of the cow's udder, squirting a stream of milk at the calf's mouth. The calf will lick and taste the milk, contemplating its deliciousness. The calf can then be drawn to the cow's udder, and encouraged to nibble on the teat. The farmer can even open the calf's mouth, place a teat inside, and then close the mouth onto the teat, squeezing out milk into the calf's mouth. This should be enough to get the calf on board with the nursing program.

There will always be clueless calves, that don't quite figure it out. Of course, there also can be disinterested mothers, frequently the case with first time heifers. In these situations, the first guideline is to keep the cow and calf in proximity to one another. Do not let the cow walk off to a far part of the pasture, forgetting momentarily about her calf. Keep the two close together, and allow their natural instincts to unite them during the first hours after birth.

It can also happen that the cow, especially a heifer cow, is painfully tender and does not want to allow the calf to nurse. It is so sad to watch a clueless and curious calf being kicked away by a confused momma cow. It does happen. When it does, you will want to lead the cow into the milking stanchion, restrain one of her hind legs, and then encourage the calf to nurse. If you see a cow being mean and uncooperative like this, it is good to intervene swiftly, lest the calf becomes traumatized to the point that it may not want to try nursing again. As always, once the proper pattern is established, cow and calf come to love their routine. Once things finally click, leave the pair alone to bond with one another.

Nursing will stimulate the newborn calf to defecate, clearing its intestines of the meconium. The momma cow will lick away this meconium, right as it is eliminated by the calf. The consumption of the meconium will stimulate hormones that cause the cow's placenta to be delivered. The placenta should not be disturbed; it will be naturally eaten by the mother cow. Consuming the placenta gives the momma cow a huge boost of vitamins and nutrients, helping her system to recover from calving and shift into lactation mode. The consumption of the placenta is a harmonious and important event in the natural process of calving.

Calf Care

The first week of life is a fragile period for the newborn calf. Almost all mortality either occurs in the first week, or its foundations are laid during that time. Ensuring that the calf nurses within the first 24 hours is critical for the digestive microflora and immune system of the calf. This was discussed earlier in this book, but it merits repeating that the calf must consume at least one full cup of colostrum milk during the first day of its life, or it will never develop into a healthy animal. If necessary, you can even hand milk the colostrum, and pour it down the calf's throat, in a worst-case scenario. The importance of adequate colostrum consumption during the first 24 hours of life cannot be overstated.

The umbilical cord will have snapped off during calving, and there will be a bloody end exposed to the elements. In a clean pasture setting, particularly in an arid environment, no further attention is necessary. If you are in doubt as to the cleanliness of the calf's environment, you should dip the bloody end of the umbilical cord in a dilute solution of iodine. This is just to ensure no infection can occur in the vulnerable and exposed wound.

Vaccinations of the calf are completely at the discretion of the farmer, and depend largely on the local disease pressures that may exist. In a closed herd, in a healthy region, there is no need, nor reason, to vaccinate our calves. This is a contentious topic, and if in doubt, consult with other local farmers. Your local veterinarian will likely be pro-vaccine, and as such, not the best source for unbiased information. It is true that there would be nothing more unfortunate than losing an animal to a vaccine-preventable illness. However, vaccination, in animals as well as in humans, does carry risks, and unnecessary medical treatment always should be avoided.

Milk Fever

Milk fever is a serious condition that can be fatal in high producing dairy cows. All lactating mammals are susceptible to milk fever, but it is most problematic in dairy animals that have been selected for maximum milk production.

Milk fever is a paralysis that occurs to cows, generally one or two days after calving. Untreated, the paralysis progresses from the cows' limbs, to her lungs and heart, resulting in a fatal condition. Milk fever is an acute depletion of calcium in the cow's bloodstream, caused by the tremendous calcium demands in the udder for milk production. The symptoms of milk fever progress from loss of bodily coordination, to the cow laying down, to her being unable to stand, and finally death. The remedy for milk fever is a simple injection of soluble calcium into the cow's muscle tissue. If in any doubt as to milk fever, call a veterinarian immediately. Milk fever is easily treated. Once the cow receives the calcium injection, she will typically recover so quickly as to be standing up and walking around before the vet leaves the farm. The especially prepared dairy farmer would do well to keep several syringes of injectable calcium on-hand, in case of an emergency.

Prevention of milk fever is best accomplished by feeding the cow low nutrient feed for a month after drying up in the winter. Somehow, this period of lean feed helps the cow to better mobilize and utilize calcium in her body. Synergistically, this important period of low nutrient feed happens to be the very best thing for a dairy cow that is being dried up. This is why it is good holistic cow management to have our cows eating the last remnants of poor pasture during the month after they are dried up. After this drying up period on low nutrient feed, it is good to then feed excellent alfalfa hay for the remainder of her pregnancy, until the pastures are ready for grazing to resume in the spring.

Do not feed cows that are recently dried up a diet containing a large percentage of alfalfa, as this will compromise the cow's ability to mobilize calcium in their bloodstream, making them more susceptible to milk fever with their next calf.

Milk fever has a definite genetic component to its occurrence. Frustratingly, it seems to afflict the highest producing cows. It is recommended that any cow who demonstrates susceptibility to milk fever should be culled from the herd. Left in the breeding herd, propensity to milk fever will likely be passed on to the cow's heifer calves. Through careful selection, milk fever can and should be eliminated from a healthy and holistically managed cowherd.

Special Considerations for Bull Calves

Polling all bull calves during the first week of their lives is a prudent management decision. Bulls are dangerous and destructive enough without their horns. While the horns of the cow are hollow sensory organs, filled with nerve tissue and blood, the horn of the bull is solid like a fingernail. Accordingly, we should never deprive our cows of their horns, which are mystical organs of perception. Horns on the bovine bull have no sensation or perception, and are used exclusively as weapons. Removing the horns from our bulls is a wise decision for the holistic wellbeing of our farm.

The best way to poll our bulls is with a caustic dehorning paste. It is applied with a small stick, to an area the size of a dime, right where the faint bump of the horns can be felt on the skull. Apply the dehorning paste once the calf has been separated from its momma in the evening, after the first week of them being together all the time. The calf needs to be apart from the momma when the dehorning paste is applied, or the momma will lick away the paste, defeating its purpose and potentially sickening the cow. The paste is applied in the evening, and by morning it will have dried completely. Carefully brush away any residue, and of course, follow all directions on the dehorning paste product label. The dehorning paste burns the edge margin of the growing horn, stopping its growth permanently. This one-time procedure is the only action needed to permanently poll our bull calves.

Castration is only a worthwhile procedure for animals destined to be raised for oxen. As previously discussed, intact bull beef is an excellent product on the small farm. There is no reason to castrate our bulls into steers if they are being raised for beef. Dual-purpose cattle breeds, such as are recommended for a holistic, pasture-fed dairy herd, can make excellent oxen, if raised and trained properly.

Young bulls that demonstrate excellent temperament, and good physical conformation, could be considered for working draft animals. Castration should occur at three months of age, well before the onset of testosterone development in the animal. A professional cattle vet should handle the procedure. Castrating when young causes less physical stress for the animal, resulting in fewer health repercussions.

The process of training an ox takes an enormous amount of time and dedication. It is not a task to be taken lightly, or poor results will surely be the outcome. An excellent book for those serious about raising working oxen is *Oxen: A Teamster's Guide* by Drew Conroy. Additionally, the excellent publication *Small Farmer's Journal*, published by Lynn Miller, contains many contemporary articles on raising oxen on small farms.

All bulls that are to be raised for breeding stock must be nose rung by one year of age, preferably around the six-month mark. A mature bull will be completely unrestrainable without a nose ring. In older times, when bulls were common on small farms, there were laws prohibiting keeping a bull past one year of age without a nose ring, because if he escaped, his capture would be both dangerous and difficult. With a nose ring, even a large and uncooperative bull can be safely transported or restrained, in a humane and painless way. Nose rings are absolutely essential for mature breeding bulls.

The easiest way to ring a bull is to sedate the bull unconscious with a tranquilizing injection. Some farmers will ring their bulls using a head catch and squeeze chute, but this seems both cruel and physically challenging. A tranquilizing shot of Ketamine, administered by a veterinarian, will temporarily sedate your bull while you ring him painlessly. A special trocar knife is used to puncture the nasal septum, and the nose ring is then inserted into the opening. This procedure is best learned hands-on from a skilled cattleman or vet. The nose ring is secured permanently in the bull's nose; it will not be outgrown or ever need replacement.

A bull can be caught by his ring, with a long metal pole with a hook on the end. Particularly when the bull is distracted eating hay, the hook can discreetly be slid into position, and hooked securely. The bull is then held in position, and a carabineer attached to a long rope is clipped to the nose ring. The bull can now be lead securely, though extreme caution is still needed when handling a bull on a rope. A creative thinking exercise where you imagine the infinite number of ways that the farmer could still get severely injured in this situation will help to ensure that these tragedies never befall you.

When grazing a bull on a tether, tie off a rope directly to the nose ring. This is to prevent the metal carabineer from wearing away the metal on the nose ring. Any rope used to handle your bull should be very strong, with an extremely durable exterior sheath to resist abrasion. Climbing ropes have proven very effective for their ease of knot tying and durability of sheath. Be sure to always utilize secure rope work with redundant knots when handling your bull.

Training your bull to be comfortable on rope begins at a young age. A broad understanding of the concepts of Natural Horsemanship will be helpful for working with your bull. Mark Rashid is a highly recommended author on the subject. Understanding the mindset of your animal, and the signs that animals give, will make training a much more successful endeavor. Above all else, always place safety first in your dealings with the bull.

12. MILK SALES AND CUSTOMER RELATIONS

Selling our Raw Milk

The sale of raw milk is regulated on a state-by-state basis, with widely varying laws across different states. Researching the specific legal situation in your home state is absolutely necessary to ensure smooth operation of your dairy business. Two non-profits, the Farm-to-Consumer Legal Defense Fund, and the Weston A. Price Foundation are excellent national level resources for information on raw milk laws. Consultation with them, and your state department of agriculture, are recommended steps in the early planning stages of your raw milk dairy.

Raw milk legality generally falls into five broad categories for individual states. These categories are retail sales, farm direct sales, herd shares, pet food, and illegal. State laws are frequently changing, generally towards a more permissive environment for raw milk access for consumers. Even if your state laws are problematic now, they may well change for the better, and an engaged citizenry will help to move raw milk laws in a more sensible direction.

In a few states, retail sales of raw milk are permitted. This means that you may be able to sell your raw milk at a local grocery store, though frequently there will be significant state regulation of production facilities in these circumstances. Retail sales generally create a problematic regulatory environment for small farm dairy producers. Legal consultation regarding the production stipulations placed on raw milk producers in these states is imperative.

Farm direct sales mean that consumers are allowed to purchase raw milk, but the sale must be direct from the producing farm. This is a good arrangement that typically allows relative autonomy for the producer, with the consumer responsible for purchasing milk from a local farmer they know and trust. Being located close to your customer base will be essential for a farm direct raw milk dairy to be successful.

Herd shares are a legal arrangement that allows for consumer access to raw milk without technically sanctioning the sale of raw milk as an agricultural product. This situation may seem complicated, but as will be discussed in detail below, herd shares are actually the best arrangement for the dairy farmer. Herd shares create a dedicated customer base, and provide legal protection to the farmer. Even if you live in a state that allows retail or farm direct sales, herd shares provide many advantages to small dairy farmers.

Pet food sales are a condescending legal arrangement that refuses to acknowledge the healthiness of raw milk, while still allowing consumer access. It is an example of a paralyzed government turning a blind eye to something they do not feel comfortable with addressing outright. Careful assessment of the specific state statutes will allow the dairy farmer to find the appropriate legal loopholes that allow provision of raw milk to interested customers in pet food states. Hopefully these states will move to a more respectable legal framework for raw milk in the near future.

Raw milk remains plainly illegal in several states at the time of publication of this book. Needless to say, this is a tragic situation that advocacy groups such as the Farm-to-Consumer Legal Defense Fund and the Weston A. Price Foundation are striving to correct. In many states that prohibit raw milk sales, there are large industrial dairy lobbies that work to block customer access to small farm raw milk. Direct citizen action is always the best way to change problematic legislation. Investing the necessary capitol into starting a raw milk dairy in this hostile legal environment is not a good idea, generally.

Raw Milk Shares

A Raw Milk Share is a simple business agreement between farmer and consumer. Practically speaking, the system works much like a club membership. The one-time herdshare purchase is like buying a membership to join the club; the monthly boarding fee is like the monthly dues.

The Raw Milk Share system is similar to a traditional farm Community Supported Agriculture membership. CSA's are based on a financial commitment from the customer to receive product on a regular and recurring basis. This guarantee is good for the farmer, who will be milking and feeding the cows daily, regardless. The similarity to the CSA model gives the farmer a sure income stream, and gives the customer a certain supply of farm fresh raw milk.

The relationship begins with the farmer selling the customer a 'share' in their cowherd. The customer now technically owns a small fraction of the cows in that herd. The purchase of a share in the cowherd is a one-time expense, good for life. Once a shareholder, the customer then pays for the daily care of their share of the cowherd; this is the monthly 'boarding fee'. As a dividend of ownership, the customer receives milk, from their cow, on a weekly basis. The customer is billed monthly, with the amount due depending upon the agreed upon amount of milk they receive each week.

The Raw Milk Share model includes a formal contract that describes the customer's monthly boarding fee. In this way, it is no different than boarding a horse at the stable. The customer pays for the regular care of their cow, not the milk itself. If a customer misses their milk pickup, they are still responsible for their full monthly boarding fee. The fresh milk is considered a dividend that the customer receives, as a byproduct of caring for their share in the cowherd. Sometimes it takes a little explaining for the customer to understand that they are not actually buying raw milk. Ultimately, the system is straightforward, and is modeled on other common business arrangements with which the customer is already familiar.

The milk share system remains flexible, so that if needed, customers can change their subscription level on a monthly basis. During the winter dry period, when no milk is available, the customer is not charged anything. When the cows freshen in the spring, the milk share program starts back up again, monthly dues resume, and milk is once again provided on a weekly basis. The Raw Milk Share program works seamlessly throughout the year.

Setting Prices

Determining the ideal price point for your milk will be essential for the profitability of your small dairy business. The specific pricing structure that you establish will be highly dependent on local economic factors. Surveying other regional producers will give you some guidelines for pricing. Remember, just because another dairy charges a given price, it does not mean that they are running a viable business. This is particularly true with startup farms charging low prices. They may well be pricing themselves out of business, and you do not want to follow their lead off a financial cliff.

Low prices are the marketing tactic of large agribusiness. The small farmer cannot succeed in the marketplace with this outsized business mentality. The marketing angle of successful small farms is quality and uniqueness. You are much better off charging a higher price than your established competition, rather than thinking that a lower price is needed to attract customers. The objective is to combine a higher price, with better consumer value, by providing excellent customer service and superior product. Remember that your profit margin is generally the last twenty to forty percent of your retail price. When you lower your price by fifteen percent, you have cut your profits by as much as half.

In a raw milk share program, prices need to be determined for both the initial herd share, and the monthly boarding fee. The herd share, being a one-time expense, should not be a barrier to new customers. In many ways, the herd share is a legal formality, and an act of faith on the part of new customers. It is our monthly boarding fees that constitute the primary income for our dairy operation. Keeping the initial cost to our customers reasonable encourages new business. If we were to double our herdshare fee, it would have a negligible effect on our long-term bottom line. It would, however, create a barrier to new customers joining our raw milk program. It is therefore recommended to keep your share fee modest. The share fee should be just enough to ensure that you are attracting serious customers.

Unique product offerings are the real marketing strength of small farms. When given a chance to offer a conventionally available product, or a specialty artisanal product, the choice is easy. Products like raw milk and raw milk yogurt create their own interest. You are no longer directly competing with the supermarket and huge agribusiness corporations. Once customers have the opportunity to access such artisanal products, they will be loyal to your farm for years to come. Offering unique products helps to attract attention and new customers, and to retain those customers for years.

The Legal Documents

The legal framework of raw milk shares is born from the world of lawyers and legislators. Fortunately, there are pre-drafted 'boilerplate' contract documents that you can get from the Farm-to-Consumer Legal Defense fund, or your state raw milk producers association. The key pieces of paper are the "Bill of Sale", the "Boarding Contract", and the "Production Guidelines". With these form documents, the farmer and consumer merely 'sign on the dotted line'. To actually read and comprehend these legal documents would require both a law degree and a fine sense of high old English. Customers generally get a laugh when reading the convoluted language of the contract. Lawyers take pride in actually understanding the archaic terminology; the rest of us are glad that we don't.

Every customer receives a formal "Bill of Sale", much the same as a title to a house or a car. This is their documentation that they are legally part owners of the cowherd. The "Boarding Contract" articulates the exact terms of monthly payment and the dividend of weekly raw milk. These documents, abstract as they may seem, are actual legal documents, which establish binding commitments for both parties. As such, the use of legally vetted, 'boilerplate' documents is strongly encouraged.

The third and final document for defining your Raw Milk Share Program is called the "Production Guidelines". Each customer receives a detailed written description that outlines the specific production guidelines that the farmer pledges to follow in the care of the cowherd and the production of the raw milk. Given the public perception concerns surrounding raw milk, this clear description of farming practices is very good for both farmer and consumer.

The farmer has a contractual commitment to follow these production guidelines. The language of the boarding contract legally binds the farmer to following these production guidelines. The consumer gains a clear understanding of the practices that will be followed to ensure the highest quality raw milk. Significantly, in the litigious times that we live, the farmer is absolved from legal responsibility for any health concerns that may come about from the consumption of the raw milk, *so long as the specific terms of the production guidelines have been followed.*

This legal protection is critically important, because if a customer gets food poisoning for any reason, their doctor will likely assume that their consumption of raw milk was the cause. With raw milk and the conventional medical system, there is implicit guilt by association. Because we know that properly produced raw milk will not cause anything but good health, such ignorance on the part of the conventional medical community is highly frustrating.

The terms of the boarding contract protect the dairy farmer from misguided or unscrupulous legal prosecution. If the farmer is negligent in his implementation of the agreed production guidelines, then he has broken his end of the contract, and could be found legally liable. So long as the farmer follows the practices that have been agreed upon in the Production Guidelines, he can operate his dairy farm with no fear of litigation.

Billing and Bookkeeping

Efficient billing is essential for time management, especially during the busy farm season. I often tell customers, "I am a great farmer, but not that good of an accountant." When I am selling vegetables or fruit, it is simple; I include an invoice with the produce, and check off the invoice when I am paid. For our raw milk shares, we invoice monthly, but frequently customers will pay for several months or a whole season at a time, or forget to pay for a month. In any event, good record keeping is essential to minimizing the time you spend bookkeeping.

Offering a small discount for full payment at the start of the milking season is a good strategy. Many customers will not be able to afford this large, upfront expense. But for those who can, it will minimize your accounting, and it is worth it to incentivize this option for your customers.

Setting up an automated online bill pay system for your raw milk program is a good idea. Allowing a computer program to track your customers and their payments saves the farmer valuable time and hassle.

Customers that want to frequently change their amount of milk received are a real burden on your bookkeeping. Only allow changes to milk orders at the start of the month, with prior notice, so that billing can function smoothly in monthly increments. We most always welcome increases in milk orders, but if a customer is frequently increasing and decreasing their orders, do not hesitate to explain to them that this is a real hardship for your business.

The customer and farmer relationship is inherently close when running a raw milk share program. Your customers should understand that a family dairy is not the same as a full service supermarket. Good communication will ensure that your customers work to keep your accounting burden to a minimum, so that you can focus on being the best dairy farmer possible.

On Farm Milk Pickup

Once a week, your customers will need to get their fresh milk. The two primary options you will have, both of which have their own sets of pros and cons, are customers picking up milk directly from the farm, or you delivering milk to your customers. Remember that you will be dealing with a potentially large volume of milk each week. This could entail dozens of individual customers coming to the farm, or delivery of many, many glass jars of fresh milk. It should be noted that some states will regulate your choice on this matter, and may require direct on-farm customer pickup for raw milk. Generally, these regulations can be worked around, and various systems of cooperative milk delivery can be arranged. What follows are some considerations to help you decided how best to distribute your raw milk on a weekly basis.

Direct on-farm pickup will work best when your farm has good vehicle access. One aspect of this is your proximity to your local population centers. America is a culture of driving, and customers likely will not mind a ten or fifteen minute drive out to the farm each week. For the truly dedicated customer, they may be willing to drive as much as an hour to your farm. The further you are located from town, the harder it will be to retain customers. More important than your actual mileage from town, is the quality of your access roads.

The tricky part of milk pickup often begins once the customer gets to the farm gate. Most on-farm roads will be gravel, at best. They may be fine for you and your farm family to come and go, but what happens when you have dozens of vehicles driving onto the farm, often with rain or snow, each week? Is there adequate parking and turnaround space for city people who might not be comfortable with muddy parking areas? The volume of customers creates the challenge for on-farm pickup. Many farms have access driveways that work fine for light use, but become overwhelmed and degraded quickly when a large number of vehicles are entering and exiting on a nearly daily basis.

When customers come onto the farm, there is an issue of legal liability that must be considered. If a car gets driven into a ditch, or worse, into a barn, you will be glad that you had a farm liability insurance policy that covers these types of accidents. Personal injuries could potentially happen as well, when you have dozens of individuals coming to the farm every week, there is quite a significant exposure to risk that the farmer faces. Certainly you hope for the best, and manage your farm to limit risks to your customers. But accidents are called accidents for a reason, and if customers will be coming onto the farm on a regular basis, you should be sure to find an appropriate farm liability insurance policy to protect you when an accident inevitably occurs.

Setting hours for customer pickup will be another consideration. You will need to have defined pickup times, so that you can guarantee that the milk is ready when customers arrive. You will want to have a closing hour as well, so that customers are not pulling into the farm in the wee hours of the night, disturbing your family and livestock. Establishing clear hours for milk pickup will be necessary for both you and your customers.

When your customers come direct to the farm for milk pickup, they could potentially fill their own jars from the bulk milk tank, but this comes with some significant risks. The dairy building, where your bulk tank is located, is a sanctuary on the dairy farm. Allowing dozens of individuals, with muddy boots, dirty hands, and trailing children, to enter your dairy sanctuary may not be desirable. Having customers fill their own containers is nice, because then you do not have to concern yourself with cleaning each customer's jars, as they are responsible for their own jars. But the risks of customers inadvertently doing something wrong to the bulk tank, or disturbing something of your dairy equipment, are worrisome.

A large refrigerator, in an adjacent room to the dairy, provides a good solution to the above concerns. You will bottle out the milk yourself, and leave it in the refrigerator for pickup. Your customers will leave last week's empty jars on a shelf, and pickup their freshly filled jar from the refrigerator. The key considerations in this system are ease of vehicle access, and safety of farm operations. Never underestimate the problems that town people will find on a farm. Surprise rooster attacks, overfriendly calves, slipping on ice or mud; these are all legitimate concerns when customers are coming onto the farm in an unsupervised manner on a regular basis.

Off Farm Milk Delivery

Delivering your milk may be a better option, unless the circumstances of your farm are perfect for direct customer pickup. Your pickup location will need to have a dedicated refrigerator for your raw milk deliveries. This refrigerator may be something that you purchase used at a good price, and simply loan out to your pickup location. Despite the refrigerator not being located on your farm, you will be responsible for its cleanliness, and will need to make orderly appearances a priority in your milk pickup area.

Your off farm pickup location could be a back room at a local grocery co-op, a customer's house with space in a detached garage, or a local business who appreciates the increased traffic flow of your weekly customers. All of these situations can work out well. A central location that is convenient to both you and your customer base is the primary consideration. Adequate parking and access, along with good lighting, are other factors that determine the suitability of a given pickup location.

Off-site pickup will require you to build relationships with the owner of your pickup site. You will want to find a mutually agreeable way to compensate this person for the hassle of hosting a milk pickup site. Frequently, this can be as simple as providing your host with some extra dairy product as a gift each week. There will be a large traffic flow of customers, and you will need to mediate between your pickup site owners and raw milk customers, to make sure that everyone stays happy with the arrangement. Losing a pickup site mid-season due to poor communication is a serious risk, and a real fiasco for your dairy business. Working hard to maintain a solid relationship will ensure that your milk pickup site can be as reliable as your milk cows.

Transporting your milk from the farm to your delivery hub represents a significant undertaking of its own. Assuming you are delivering in glass jars, precautions need to be taken to avoid breakage. The best system for transporting glass milk jars is to use coolers with pieces of cardboard tucked between each glass jar. A snug fit prevents jostling, which will easily cause breakage. Keeping your jars contained within a cooler is good for the inevitable broken jar that will occur, despite your best efforts to prevent an accident. It only takes one broken jar of milk, soaking into the back seat of your car, and stinking for weeks, for you to realize the importance of careful handling. A fleet of plastic coolers, secured in the back of a pickup truck, is the best way to transport large quantities of milk for customer delivery.

Glass Jars, Metal Totes, and Plastic Jugs

The half-gallon, glass canning jar is the best container for transporting milk to your customers. Glass jars filled with fresh raw milk visually demonstrate the quality and uniqueness of your product. Customers are nostalgic for the days of glass milk jars delivered by the milkman. We can nourish this dream, by providing it to our customers each week. Customers will gladly pay more for a product that is so special, and expresses such quality. As discussed previously, half-gallon canning jars are best purchased by the case, in wholesale quantity from a local grocery store. Gallon size glass containers are too breakable, and not available as economically.

As the operator of a raw milk dairy, you will end up being the owner of quite the bounty of half-gallon glass jars. The cost may be reasonable individually, but by the hundred, it definitely adds up. For every gallon that you have going to a customer weekly, you will need four glass jars. Two jars for this week, and two that the customer will return from the previous week. You will inevitably need extras on hand, as somebody will forget to bring back their glass jars when they come for milk pickup. A checklist, at the pickup refrigerator, lets customers indicate that they have picked up this week's milk, and returned last week's jars.

The best way to ensure that your customers value their glass jars, is to charge a small annual glass jar fee. We charge each customer $10 per year for glass jars, regardless of how much milk they get weekly. When customers pay for something, they value it more. Free makes people take it for granted. The $10 jar fee reminds the customers how special it is to be getting their milk in glass containers. The small fee covers the farmer for the inevitable breakage and unreturned jars. The combination of a weekly checklist, and an annual jar fee, ensures that we are not losing money on glass jars. The customers are appreciative for the uniqueness of receiving milk in glass jars, and the farmer can focus on running a dairy rather than counting and chasing missing glass jars.

Stainless steel milk totes are a good way to deliver larger quantities of milk. We have milk cans that are three gallon and five gallon. The benefit of washing and delivering one, five-gallon can, as opposed to ten individual glass jars, is self-evident. Of course, purchasing the stainless steel milk totes is expensive. You can encourage your customers to purchase their own milk totes that they own, and you simply fill with milk each week. These customers would not have to pay the annual glass jar fee. Stainless steel milk totes will not be a viable alternative for most customers, but it can be a good option, and is worth keeping in mind as a possible solution. It is particularly viable for customers who get a large quantity of milk each week for their own home dairy processing. Stainless steel milk totes can be purchased from Hamby Dairy Supply in Missouri, and Hoeggar Supply in Georgia.

Plastic milk jugs are the final option for milk delivery, and truly an option of last recourse. Plastic leaches chemicals into milk; this is a scientifically proven fact. Bottling our top quality raw milk into a contaminating plastic container is a minor tragedy. Whereas glass containers radiate quality, cheap plastic degrades the value of our milk in the customer's mind. The only reason for considering plastic milk jugs is if your local road conditions are so poor that transporting glass jars is simply not feasible. In any case, try to find a way to make it work with glass. Glass milk jars embody the exceptional quality of our farm fresh raw milk, and represent your business in the best possible light.

Testing your Milk

Sending samples of your milk to a laboratory for testing provides customers an extra assurance regarding the healthfulness of your milk. We can utilize the California mastitis test kit in the milk stanchion, as the first indicator that there may be something less than ideal about our milk quality. However, this method only alerts us to elevated bacteria levels in our raw milk. In order to know if there may be specific pathogens in our milk, it must be sent off to a laboratory for testing. Your state raw milk producers association, or the Farm-to-Consumer Legal Defense fund will be good resources for locating a local testing laboratory. In the early days of your raw milk dairy, when uncertainty about your cows' health history may be a concern, weekly or monthly milk testing can provide a significant confidence boost to the dairy producer.

Testing our milk on a regular schedule is not a perfect solution. It is far more important that we care for our cows in a manner that makes pathogenic illness a highly improbable occurrence. The primary shortcoming to milk testing as a product safety measure is the inherent delay between testing and results. Laboratory turnaround time, including shipping delays, mean that by the time we get results back, several days or a week will have elapsed. We do not get to learn if today's milk is healthy, rather we only learn that our milk was problematic in the past. Often times, with the schedule of milk delivery, we will learn that there was a problem for milk that already reached the customer for consumption. Certainly, it is valuable to know if we have any quality issues with out milk. It should be emphasized however, that proper care of our milk cows, to prevent health problems in the first place, is a much better customer safeguard than regular milk testing.

The upcoming chapter on holistic herd care will discuss in depth the important management tactics that the dairy farmer should use to ensure optimal health in the herd. For now, let us emphasize that natural diet, seasonal management, and maintaining a closed herd are the three best insurance policies against health problems.

State and Federal Milk Laws

For the past half-century, sales of raw milk have been severely restricted in most states. There is a substantial history behind this legislative repression of farm fresh raw milk. Fortunately, in the past decade, on a state-by-state basis, new legal frameworks have emerged for the legal sales of raw milk. Driven by consumer demand, more and more states are allowing consumers direct access to local farms in order to procure fresh, raw milk.

The most important consideration for a small dairy wanting to begin selling raw milk is that the laws governing raw milk sales are state laws. As such, each state has different regulations and restrictions on how raw milk can be provided to the customer. It is imperative that each farmer thoroughly research their state's specific regulations with regards to raw milk. Other than a few backwards states that still prohibit raw milk entirely, there are three basic ways in which consumers are allowed legal access to raw milk. These are retail sales, farm direct sales, and raw milk share programs.

Retail sales and farm direct sales are established methods for farmers to sell their products; they do not need much discussion here. The uniqueness of raw milk share programs, and the specific legal protections they offers small farmers, is the reason this book focuses on milk shares as the primary means of making small dairy work for small farms. Even in states that legally allow retail or farm direct sales of raw milk, herd share programs should be considered as the most secure and profitable means for small farmers to sell their raw milk.

The USDA, the branch of our federal government that regulates agricultural production, has no jurisdiction over small dairy production of raw milk. This is the reason we concern ourselves with state regulations exclusively. Commercial dairies, which rely upon pasteurization of their product, are regulated by the USDA under the Pasteurized Milk Ordinance, or PMO. Fortunately, as a raw milk dairy, you are not subject to any of the regulation contained in the PMO. These details may seem trivial, and for the most part they are. Nevertheless, knowing the law as a farm producer is essential to maintaining your freedom to farm good food.

Media and Public Perception

Promoted by the multi-billion dollar corporate dairy industry, there has been an extensive campaign of fear and misinformation over the past half-century to discredit raw milk as a safe and wholesome food. When operating a raw milk dairy, it is important to understand the powerful effect this propaganda has had on the American consumer. Raw milk has caused far fewer cases of food-borne illness than pasteurized milk, yet there remains a perception that raw milk is a risky food product that is prone to contamination. As we begin operation of our raw milk dairy, we will frequently need to educate our potential consumers about the excellent safety record of raw milk. Fortunately, the facts are on our side. Unfortunately, history and the loud platform of the conventional media are not. As producers of raw milk, it is important that we provide good information to our consumers so that they can make educated choices about consuming raw milk.

Raw milk, it should be remembered, was the only milk that our ancestors drank for centuries. Raw milk was seen as a food of great health and nourishment. Problems began to arise in the late nineteenth century when factory farms came into existence in urban environments. Conditions of animal care and milk handling were frequently disgusting, with minimal guidelines for sanitary dairy operation. As consumer protection laws began to be established, commercial dairies faced understandable scrutiny for their unhygienic practices. Unfortunately, the cause of the health concerns was twisted for corporate gain.

The dairy industry worked to convince lawmakers that the problem was a lack of industrial processing, rather than their disgusting factory farming practices. The beneficiaries of this campaign of manipulation were large dairies, who could afford the expensive pasteurization machines that they advocated. Industrial dairies were able to cynically push small family farms out of business. Laws intended to protect consumer health were ultimately subverted to increase corporate market share. The American consumer got the short end of the stick, while the dairy industry enriched itself.

The healthfulness of raw milk was well known to doctors at the time. Back in the early 1900's, doctors in New York State developed their own production protocol for raw milk, to ensure a quality supply of raw milk for patients in the New York hospital system. These guidelines, which form much of the basis for our modern raw milk production standards, ensured that raw milk would be produced to hospital standards. Such healthful raw milk was considered a healing medicine by doctors, who prescribed it to convalescing patients recovering in the hospital. This is the real story of raw milk, a healing food that should be produced in conditions that protect and guarantee its nourishing qualities.

Nowadays, there is much controversy surrounding raw milk. Many holistic medical doctors promote the consumption of raw milk for its benefits on the digestive and immune systems. Sadly, there are always detractors, frequently funded by the pharmaceutical and agribusiness industries, who loudly dismiss any such claims. Fear is used as a weapon to make parents doubt the safety of raw milk as a food for their children. Misinformation abounds, promoted loudly by those with ulterior, financial interests. This is the context in which we produce and sell our raw milk.

We small farmers do not have the billion dollar PR budgets to sway public opinion overnight. We must remain calm and patient, confident and educated, doing the good work of small dairy farming. Our customers will be our greatest advocates, as they experience first-hand the health benefits of drinking farm fresh raw milk.

Working with Your State Regulators

It bears repeating that the regulations surrounding the production and sales of raw milk exist at the state level. The USDA does not have any jurisdiction over our dairy operations, when we are providing unpasteurized milk to our customers. State laws on raw milk are changing rapidly across the country. These changes are consistently making access for consumers to raw milk much easier. The trend is towards greater availability of raw milk, with more opportunities for small farmers.

As new state laws change the legal realities of raw milk production, local bureaucrats can be slow to respond. It is important to realize that the same state government agencies and agents who once enforced laws prohibiting raw milk, are now tasked with implementing laws that allow raw milk. There can be reluctance and resistance from these individuals and their institutions, thus it is all the more important that we clearly understand the laws governing raw milk in our state.

In my state of Colorado, in the immediate wake of the passage of the Colorado Raw Milk Shares Statute, I contacted the State Dept of Public Health who was newly tasked with the oversight of raw milk dairies. The gentleman in Denver told me that they were not comfortable with the new Colorado Raw Milk Shares Statute, and had no plans to be involved with oversight of raw milk sales in the state. This abdication of governmental responsibility is a consequence of old players unwilling to adapt to a new system of rules. In this type of situation, it is particularly important to know the laws in your state, as your government regulatory agencies may not be as current as would seem fitting.

Customer Relations

The relationship between farmers and their raw milk share customers is naturally a more intimate relationship than between typical sellers and buyers. We can think of a Raw Milk Shares Program as a modified co-op, where cooperation is an essential part of the arrangement. Both parties are entering into a formal legal agreement, which necessitates good communication and faithful action on both sides. Prospective small dairy farmers must understand that more time will be invested in communication with customers regarding the raw milk dairy, than for traditional farm crops, like vegetables or eggs.

Treating all of our customers fairly is important for our public relations. Often, we try to do a favor for a customer, offering them something at a discount or going out of our way to help them with their milk pickup. Experience has shown that this almost always does more harm than good. Word of mouth spreads quickly, and when other customers realize that somebody else got a special deal, hard feelings are sure to develop. Suddenly other customers will expect similar favors to be done for them. The situation can quickly become an out of control slippery slope, where customers develop hard feelings because they feel that others are receiving preferential treatment. The solution is simple; treat all customers equally. Exceptions are best avoided, so that all customers feel equally valued. Having clear guidelines for payment, pickup, and process ensures that everyone feels equally positive towards your farm.

Developing networks with other local farms is a great way to promote and enrich your business. Collaboration with other farms, to provide milk to their established meat or vegetable customers, benefits everyone. Given the fluctuating volume of milk due to the seasonal production curve, trading seasonal milk surpluses for other farm products, such as summer fruit, is a mutually beneficial arrangement. Learn to see your fellow local farmers as opportunities rather than competitors; you have little to lose and much to gain.

Farm Tours

Inviting customers to the farm will build personal relationships that strengthen customer loyalty. As small dairy farmers, we should take pride in the work we do and the way that we manage our farms. Brining customers out to the farm will help to bring them into the life of the farm, where they will understand and appreciate more of what we do. In an ideal world, all new customers would come to the farm, see the cows and the facilities in person, and discuss the details of the legal documents they are signing. Such a policy of active and engaged communication goes a long way towards ensuring a positive relationship of trust and gratitude between farmer and customer.

Remember that our customers will inevitably have skeptical friends and family members, who may not be so educated on the benefits of raw milk. When our customers are questioned about their raw milk, we want them to be able to explain the wonders of raw milk and the beauty of the farm it comes from. In this way, our customers can help to educate future customers. Word of mouth advertising will be our best method of attracting new customers. There is no better promotion for the farm than the endorsement of an enthusiastic and knowledgeable customer.

When welcoming customers or guests onto the farm, it is important to remember to see the farm through their eyes. As farmers, we are accustomed to the sights and smells of a real, live farm. Our customers might be a bit more, shall we say, sensitive to compost heaps, or piles of feathers from a recent chicken butchering. This is not to say that we should attempt to make our farm into an artificial fantasy; we certainly should not. But we should remember that when customers come onto the farm, we are responsible for their perception of what they experience.

Do take a few minutes to tidy up. When people see a neat and organized farm, they assume that we are orderly and professional farmers. Explain to your customers what they are seeing, smelling, and experiencing. Define the narrative for them, rather than letting their uncertainty tell them stories. Above all, it is important to recognize that people's experiences on our farm will leave a lasting impression. Professionalism in appearance and explanation will yield great dividends in trust, confidence, and future business.

Owners of shares in our cowherd deserve special treatment when it comes to farm tours. These individuals have a financial commitment to, and a vested interest in, our farm. They have already paid into our farm. It is part of our side of the agreement to enable them to visit the farm and see our dairy operations, first hand. For members of the general public, who are interested in visiting the farm for a tour, the relationship is different. I absolutely recommend charging people other than your dairy members, for farm tours.

Farm tours have great value for members of the public, who often do not get the opportunity to visit a real working farm in all its beauty and wonder. Farm tours also occupy meaningful hours for a busy farmer during the growing season. Consider the value of your time, and also consider the cost for similar recreational or educational activities for the general public. Do not be afraid to charge a fair price. Your tour participants will value their experience more if they paid a fair price for it. You will be more eager and enthusiastic to give excellent quality farm tours if you are fairly compensated for your time and knowledge.

Selling Animals from the Herd

The annual process of breeding and calving produces a large number of new offspring on a yearly basis. The home dairy herd is a more than renewable resource; it is a wonderfully self-generating resource. Soon, based on the limited pasture on the homestead, and finite number of customers, selling animals from the herd will become a necessity. Managed properly, the sale of surplus animals from the herd provides a significant addition to our farm income, while improving the genetic quality of our herd in the process.

A heifer calf is always a fortunate addition to the dairy farm. From the day of their birth, we will be watching and evaluating our heifers, hoping to see the best traits emerging. In time, we may notice something that makes us decide that a given heifer does not have a future in our milking herd. The specific attributes we are looking for are discussed in detail in the chapter on cow selection and genetics. For now, let us just consider what to do with a heifer that we decide will not become a future part of our milking herd.

The best time to sell a heifer is once she has been bred, and we are confident that she is pregnant. At this point, our heifer will be two and a half years old, and almost full size. We will have invested a lot of time and expense into her care. We will know her well, and hopefully be fond of her. Although we have decided that she is not good enough to join the milking herd, assuming she is sound in health, a bred heifer is worth a good sum of money.

Our established reputation as the operator of a quality dairy farm should attract potential buyers to us on a regular basis. Keep a list of potential buyers. When the time comes to sell a heifer, this list will provide us with direct access to interested parties. Determining the price for a bred heifer is not necessarily straightforward. Comparisons to market beef prices are essentially irrelevant. You will need to know your market, by talking with other organic dairy producers. Undervaluing a bred heifer that we have raised carefully for over two years is a real financial tragedy for the dairy farmer. Determine a fair market value, and be confident asking this amount. Haggling over price is for used car salesmen, not dairy farmers producing excellence. Take pride in your farm, know your animal's value, and be firm in your resolve to sell at that fair price.

Vetting any potential buyer is a wise precaution, especially when selling an animal locally. You can be sure that if anything goes wrong with the heifer, you will be receiving a less than happy phone call. Sometimes it makes sense, for your personal wellbeing, to take a slightly lower price from a buyer you feel confident about. It is a satisfying feeling to know that an animal that your bred and raised, is now living a good life on the local farm of a friend or neighbor you respect and trust.

As you raise your heifers, you will eventually come to a point where you have a heifer that is better than one of your established milk cows. Typically, you will not know this until the heifer has calved, and been milked for at least half a lactation. You should always be cautious to be sure that your prospective heifer is in fact a better animal than the cow she will replace in your herd. Once you sell that cow, you cannot undo the decision. But if you have milked your heifer for some months, and feel confident that she is a superior animal, then the time has come to sell one of your cows. The selling process is the same as for a heifer, though you will likely find even stronger demand for a trained dairy cow that is currently milking. A trained and proven dairy cow should command a higher price than an untested heifer. Breeding the cow before you sell her will increase her value considerably, and is always a recommended practice.

Dairy bulls have limited marketability, but if you have a top bull of breeding quality, he certainly has value to the right buyer. Similarly, if you have a mature breeding bull that has come to his end of service for your herd, he may be highly desirable to another farmer. Anything but the best bulls should be harvested for beef, or simply sold at the sale barn to get rid of them at the earliest convenience. Maintaining a network of contacts among regional dairy farmers can give you a better chance for finding a good buyer for a top dairy bull. The price for a bull is even more subjective than for a cow or heifer. A truly top class bull is worth every penny of his price, considering the profound improvement he will exert over his many offspring.

13. THE BIG PICTURE: BELLA FARM

The toolbox is complete. The pieces are all in place. We now have a system for how to operate a small, holistically managed dairy farm, in a way that benefits both farm and farmer. The dairy farm dream is ready to become a reality!

Bella Farm: A Success Story

Everything described in this book comes out of the experience of establishing and operating Bella Farm, a family-run, small-scale, raw milk dairy in Western Colorado.

One day, driving through the rural countryside of our North Fork Valley, the weight of history bore down upon me. Looking out over fertile agrarian pastures, the stories of the old-timers came to life in my imagination. I envisioned the dozens of small dairy farms that once dotted this landscape. Back when family farms were the economic backbone of this rural county, many farmers milked a few cows, and sold their milk to local creameries. The lush irrigated farmland was perfectly suited to converting grass into milk. The only thing missing was the farmers.

Overzealous regulation had pushed these small dairy farmers off the land. Ranches consolidated into larger operating units, and become devoted to conventional hay and beef businesses. The land had declined in its productive value, with many farmers disenfranchised from their agrarian heritage as a result.

Just a few years previous to my epiphany, a monumental change had occurred in the legal landscape of dairy farming in Colorado. The state legislature passed the State Raw Milk Statute, allowing direct consumer access to raw milk through herdshare agreements. Micro-scale dairies providing milk directly to their customers could operate once again. The regulatory doors had opened, enabling a repopulation of small dairy farms across the rural landscape. The time was now for small farms to reclaim their heritage of pasture-based dairy farming.

Humble Beginnings

A decision was made, and a plan was launched. Over the winter, I disassembled a large barn scheduled for demolition, and hauled it back to the farm, piece by piece. Blueprints were carefully drafted by the warmth of the woodstove. When the snows thawed, we broke ground on the dairy barn, and began construction.

Over the summer, the bulk of our farm income was carefully set aside, saving up to purchase our first dairy cows. As the dairy barn emerged triumphantly from the landscape, our savings account grew, and our plans moved towards their destiny.

The search for good dairy cows led us to good dairy farmers, both rare commodities in the twenty-first century. Many phone calls radiated outward from our little rural farm, discussing cow genetics and holistic management with any remaining dairy farmers who still farmed in the traditional ways. Leaving no stone unturned, the net was cast with a thousand mile radius, reaching out to farmers from Nebraska to Washington. The principle of learning from the best motivated my inquiries. Many pages of notes were compiled in a tattered notebook. Visits to dairy farms throughout the Western States were arranged. Great insight was gained.

Our founding cows were finally located on a Biodynamic dairy in Western Montana. These cows were managed fairly comparably to how I intended to manage my dairy herd. They came from a similar environment and climate to my own. They were production animals, not pedigreed show cows. With my savings from the summer's vegetable sales in hand, I made the long journey North, and brought back Sugaree and Vivian to Bella Farm.

I milked by hand that first winter, learning as the milk pail filled each day. Dishes were washed in a corner of the new dairy barn, with water heated on a propane burner. The nourishment of the fresh raw milk was an immediate boost to the health and vitality of my family. As the snows deepened outside, I would crouch beneath the warm belly of my cow, keeping warm through the physical exertion of hand milking.

The cows were dried off during the depths of winter, and hay was fed daily from the traditional overhead hayloft of my original dairy barn. Eventually the snows melted away just as the haystack was dwindling. Spring finally ushered in the onset of the grazing season. Our first calves were born healthy and without intervention, and the cycle of seasonal dairy farming was established at Bella Farm.

Our First Customers

That first full season of milking cows was a revelation, a realization of the value of small dairy to the family farm. We milked our two cows, chilled the milk in glass jars in a chest refrigerator, and continued saving our farm income to upgrade our dairy facilities. Our first customers were brought out to the farm, shown our operation in great detail, and signed herdshare contracts to allow them to legally receive fresh raw milk each week. We directed our income towards the purchase of professional dairy equipment, starting with a Surge belly milker machine at the start of the season. We were now milking by machine, providing milk to customers, and realizing the dream of dairy farming.

A small dairy building was constructed, as a dedicated milk handling and processing facility. On-demand hot water was installed, and we began searching for a bulk cooling tank to be able to professionally cool and bottle our milk. By the end of the first full season, these investments had been made, and our dairy became a thing of real agricultural beauty.

Growing the Herd

Buoyed by our customers' enthusiasm for our raw milk, and encouraged by our profitable business trajectory, we decided to bring in another cow, and significantly, a breeding bull. Another trip North was planned, a bull shed constructed, and the process of breeding our own cows began. The first summer we had relied upon Artificial Insemination to impregnate our cows, and after seeing the difficulty in timing pregnancies through artificial means, the need for a dairy bull was clear. Our herd was growing, our infrastructure developing, and our business began to thrive.

Our herd grew in that first year from two cows, to three cows, a bull, and two yearlings. We could see how the self-generating principle of cattle breeding would fuel the growth of our business in short order.

Within another year, as our herd naturally swelled from our annual calving, we were ready to sell a cow as our first home-born heifer began milking. This large boost of capital helped to propel our finances into positive territory, offsetting our entire annual hay bill. Our infrastructure was in place, our cowherd was delivering a surplus, and our systems of management were becoming more efficient with each passing day. The small farm dairy was proving itself a success.

Living the Dream on a Dairy Farm

The dairy herd at Bella Farm has now developed into a coherent system of holistically managed cows. The integration of our dairy herd into the larger farm system has brought abundance to every corner of the farm. Pigs and chickens are nourished by excess skim milk. The market vegetable garden and fruit orchard are fertilized by the ample compost generated by our winter hay feeding. Our pastures continue to build humus and species diversity through holistically managed rotational grazing. The cowherd is the hub around which the entire farm thrives.

Dairy cows form the backbone of our farm's identity and purpose. Our business thrives as we nurture our cowherd in accordance with natural principles of holistic management.

Our family is healthy and strong, nourished by the healthful properties of pasture-fed raw milk. Our children grow vibrantly, drinking fresh milk and raw yogurt every day. Consumption of fresh dairy products is better than the best health insurance money can buy.

The personal satisfaction gained by herding and managing these magnificent creatures, long held sacred by cultures around the world, is profound.

The images of our young children herding yearling calves into a fresh pasture are precious reminders of the beauty of life.

The majesty of a bull guiding his cows back to the barn for milking reminds us of the best qualities of natural leadership.

The tremendous power of living creation that is evidenced in each calf born on the farm continues to invoke awe and a deep reverence for life on Earth.

Sharing the Abundance

The dream is alive! The family farm thrives when organized around the historical pattern of pastured dairy farming.

I hope this book gives you the inspiration and the information needed to establish and operate your own vision of agricultural paradise. Best wishes of fortitude and fortune in all your endeavors with the sacred dairy cow.

RESOURCE BIBLIOGRAPHY

Anderson, Arthur. *Introductory Animal Husbandry.* 1943. New York, NY: Macmillan, 1951.

Ashby, Wallace, Robert Dodge, and C.K. Shedd. *Modern Farm Buildings.* Englewood Cliffs, NJ: Prentice-Hall, 1959.

De Bairacli Levy, Juliette. *The Complete Herbal Handbook for Farm and Stable.* 1952. Boston, MA: Faber and Faber, 1984.

Drayson, James. *Herd Bull Fertility.* Austin, TX: Acres U.S.A., 2003.

Fraser, Wilber. *Dairy Farming.* New York, NY: Wiley and Sons, 1930.

Gardiner, Frank. *Traditional American Farming Techniques.* 1916. Guilford, CT: The Lyons Press, 2001.

Gerish, Jim. *Management-Intensive Grazing.* Ridgeland, MS: Green Park Press, 2008.

Hand, Thos. *Guenon on Milch Cows.* 1888. New York, NY: Orange Judd Company, 1913.

Ingham, Elaine. *The Compost Tea Brewing Manual.* 5th Edition. Corvallis, OR: Soil Foodweb Inc, 2005.

Moore, Merton and E.M. Gildow. *Developing a Profitable Dairy Herd.* Hamburg, PA: Windsor Press, 1953.

Nation, Allan. *Quality Pasture*. Ridgeland, MS: Green Park Press, 2005.

Pfeiffer, Ehrenfried. *Soil Fertility*. 1947. Launceston, Cornwall, UK: The Lanthorn Press, 2004.

Salatin, Joel. *Salad Bar Beef*. Swoope, VA: Polyface, 1995.

Small Farmer's Journal. Quarterly. Lynn Miller, publisher. Sisters, OR: Small Farmer's Journal, Inc.

"The Facts About Real Raw Milk." *A Campaign for Real Milk*, Weston A. Price Foundation, 2014. Web.

Turner, Newman. *Fertility Pastures*. 1955. Austin, TX: Acres U.S.A., 2009.

---. *Herdsmanship*. 1952. Austin, TX: Acres U.S.A., 2009.

Van Loon, Dirk. *The Family Cow*. North Adams, MA: Storey Publishing, 1976.

Wightman, Tim. *Raw Milk Production Handbook*. Falls Church, VA: Farm-to-Consumer Legal Defense Fund, 2008.

Wing, Henry. *Milk and Its Products*. 1897. New York, NY: Macmillan, 1913.

ABOUT THE AUTHOR

Born in suburban Los Angeles, Adam Klaus found his passion for farming during college travels around the world. From the lush pastures of Switzerland, to the deep love of the dairy cow in India, Adam found meaning in the connection in the relationship between man and cow.

Fueled by his passion, though not coming from a farming background, Adam spent five years studying dairy farming in its most holistic, permacultural, and traditional forms.

Bella Farm was founded in 2005, as the seed for a future family farm.
Adam Klaus is a father of three homeschooled children, all born and raised in a one-room, off-grid cabin he built himself during his first year at Bella Farm.

Adam now shares his love and knowledge of dairy farming and holistic agriculture through conference speaking, educational workshops, and farm consultations.

Printed in Great Britain
by Amazon